How I Became a Quant

Insights From 25 of Wall Street's Elite

Richard R. Lindsey
Barry Schachter

John Wiley & Sons, Inc.

Published by John Wiley & Sons, Inc., Hoboken, New Jersey.
Published simultaneously in Canada.

Wiley Bicentennial Logo: Richard J. Pacifico

Library of Congress Cataloging-in-Publication Data

How I became a quant : insights from 25 of Wall Street's elite / [compiled and edited by] Richard R. Lindsey, Barry Schachter.

p. cm.

Includes index.

ISBN 978-0-470-05062-0 (cloth)

1. Quantitative analysts–United States–Biography. 2. Wall Street (New York, N.Y.)–Biography. 3. Finance–Mathematical models. 4. Finance–Computer programs. 5. Financial engineering. I. Lindsey, Richard R., 1954- II. Schachter, Barry. III. Title: Wall Street's elite.

HG172.A2H69 2007

322.63′2042–dc22

2007002383

Printed in the United States of America.

10 9 8 7 6 5 4 3 2 1

Rich would like to dedicate this book to his daughter Nancy, age 4, in the hope that she can find a partial answer contained herein when she starts to wonder what he did when she was growing up. In the same spirit, Barry would like to dedicate this book to his daughter Devra, age 18, who for many years has been asking, "What do you do at work, Daddy?", in the conviction that she will find in its contents a more satisfying answer than she has invariably received previously, specifically, "I worry about stuff."

Contents

Acknowledgments

We would like to acknowledge our editors, Bill Falloon and Emilie Herman of Wiley, for their efforts and support; Sara Pick and Paige Lesniak for administrative support throughout the process; and Rebecca Lindsey (who is not a quant!) for her careful reading and editing of this work. Barry would like to thank Karen Hoogsteen, for her always-ready support, encouragement, and patience. Finally, we want to thank each of the contributors for taking time out of their busy lives to share their experiences.

Introduction

Because you are reading this introduction, one of four things must be true. You are a quant and are intrigued by the idea of reading the stories of others like you. You are not a quant, but aspire to *quantness*, and you are seeking some insight on how to achieve that goal. You are neither a quant, nor have such aspirations, but you want to understand the way Wall Street really works, perhaps to gain some perspective on the vast and unsympathetic forces affecting your life in mysterious ways. Or, misshelved among the science fiction and fantasy titles by a harried employee, the title has struck your fancy as, perhaps, a potentially satisfying space opera. There might be other things besides these four, but we can't think of any. For all of you except the fourth group, we are pretty sure this book will provide considerable satisfaction. (For the fourth group, who knows?) By way of introduction, we will explain from our perspective the roots, roles, and contributions of the Wall Street quant.

We begin by defining the *Quant*. Mark Joshi, a famous quant, has proposed this definition:

> A quant designs and implements mathematical models for the pricing of derivatives, assessment of risk, or predicting market movements.[1]

Perhaps some of the terms used in this definition require definition themselves. A *mathematical model* is a formula, equation, group of equations, or computational algorithm that attempts to explain some type of relationship. For example, Einstein's famous $e = mc^2$ is a model that describes the relationship between energy and mass.

Quants implement models that focus on financial relationships. Perhaps the most famous of these is the Black-Scholes option pricing formula, which describes the relationship between the prices of two financial instruments that have a particular connection. The development of the Black-Scholes model (between 1969 and 1973) is often cited as one of the factors that started the quant revolution on Wall Street, but that is an oversimplification.

Returning to the definition of a quant, the *derivatives* for which quants design models are financial instruments whose values depend on (or are determined by) the future value of some quantity. This definition may seem vague—and it is. Derivatives exist in such variety that any definition hoping to be all-encompassing has to be vague.

One concrete and ubiquitous example of a derivative is an equity call option (a *call*). Someone who buys a call has purchased the future right to buy the specified company's common shares, not at the market price, but at the price stated in the option contract.

Options have been around for a long time, but one date is commonly cited as the trigger for the derivatives revolution (which is inextricably associated with the quant revolution). That date is April 26, 1973, though to call this the beginning of the derivatives revolution is an oversimplification. On this date there was an earthquake off the coast of Hawaii, but the real earthquake that day was in Chicago. The Chicago Board Options Exchange (CBOE) became the first organized exchange to have regular trading in equity options. A humble beginning, certainly, as only 911 option contracts were traded on 16 different equities. Now, each year, hundreds of millions of equity option contracts on thousands of companies trade on dozens of exchanges (both physical and electronic) around the world.

The key ingredient that ties quants to derivatives and the other two functions identified by Joshi (risk assessment and predicting markets) is mathematical know-how. The Black-Scholes option pricing formula is a good example of this.

The model, as it was first presented, was obtained by employing a result from physics, the solution to a particular partial differential equation called the *heat-transfer equation*. The level of abstractness involved in this work frequently inspires awe, fear, and even derision among nonquants. Consider this quotation from *Time* magazine of April 1994, cited by Peter Bernstein: "Prices of derivatives are not based on old-fashioned human hunches but on calculations designed and monitored by computer wizards using abstruse mathematical formulas . . . developed by so-called quants . . . "[2]

Wizards, indeed. Even Emanuel Derman, one of the most famous of quants, feels compelled to assert that "[t]he Black-Scholes model tells us, *almost miraculously*, how to manufacture an option . . . "[3] (italics added).

As the knowledge necessary to perform such feats is not a part of the regular secondary-school math curriculum, facility with derivatives requires a level of quantitative (hence "quant") training and skill confined to the mathematical specialist.

Where can these specialists be found? For Wall Street, the breeding grounds of future quants are the halls of academe, and more specifically, graduate departments of physics, mathematics, engineering, and (to a lesser extent) finance and economics. The favored candidates are holders of the degree of PhD, but not exclusively so. More recently, a new breeding ground of quants has arisen in schools that have begun teaching more focused curricula, leading to master's in quantitative finance, master's in financial engineering, master's in computational finance, and master's in mathematical finance, for example.

Okay—the reader will now be asking, "So that's what quants do and where they come from, but why do they do it?" The obvious answer, as you readers of the second group have already figured out, is that being a quant is financially rewarding. It is financially rewarding because a quant produces something with significant utility in the financial marketplace. Still, such an answer would have been greeted with derision by the famous mathematician G. H. Hardy. In his apologia to mathematics research he states,

> The "real" mathematics of the 'real' mathematicians, the mathematics of Fermat and Euler and Gauss and Abel and Riemann, is almost wholly "useless" . . . It is not possible to

> justify the life of any genuine professional mathematician on the ground of the 'utility' of his work.[4]

But before we weep for quants who, while well-rewarded financially, have failed to justify their lives when measured by Hardy's yardstick, it must be noted that the yardstick has a crack in it. The work of two of Hardy's icons, Pierre Fermat and Frederick Gauss, perhaps above all other mathematicians, have contributed to the utility of quants' efforts.

In truth, Hardy's view is really a throwback to the Middle Ages, when the idea of science was governed by the Aristotelian concept of knowledge as tautology. *In other words*, all things that we can say we know to be true can be proven true by mathematical logic alone. Utility is not a consideration. In contrast, the Enlightenment view of science (the view of Francis Bacon and his intellectual followers) defines science in terms of improving the understanding of the forces at work in the world with which we interact. In this context, utility is a natural measure of scientific contribution.

Fermat, in the seventeenth century, was the first to correctly solve certain problems related to games of chance, problems posed to him (and to Blaise Pascal, a mathematician famous for his *triangle,* among other things) by a well-known player of such games, the Chevalier de Mere, who was looking, not for Aristotelian truth, but for the proper rules to use to split the pot of cash wagered in a game that ends before there is a winner. If Fermat wasn't a quant by Joshi's definition, then we can't tell a quant from a quail. As Douglas Adams (author of the science fiction classic, *The Hitchhiker's Guide to the Galaxy*) said, "If it looks like a duck, and quacks like a duck, we have at least to consider the possibility that we have a small aquatic bird of the family anatidae on our hands."

It is worth noting that Fermat's work was not the first to address pot-splitting; methods of splitting the pot in unfinished games of chance predated Fermat's solution. However, his innovation replaced the earlier ad hoc, incorrect practice, with a new, fair distribution method.

Such is the nature of many of the contributions or innovations made by quants. Options as distinct financial instruments have been traded for hundreds of years. For example, options on agricultural commodities were traded regularly during the American Civil War. Early twentieth-century financial market participants in Chicago and New York actively traded commodity and equity options during regular time periods, albeit off the exchange floors, and prices were reported in the papers. Like Fermat, what Fischer Black and Myron Scholes (and Robert Merton) added, was a way to determine the "fair" value of an option (subject to various caveats related to the reasonableness of the model's assumptions). Once adopted, their solution replaced the prior ad hoc pricing approach.

Fermat is not the only historical example of a scientist devising a financial innovation that today would label him as a quant. A particularly striking example is the role quants played in improving government finance practices as far back as the sixteenth century. A common means of financing municipal and state debt in the Renaissance was the issuance of life annuities. In return for providing a sum to the government, the provider could designate that a regular annual payment go to a designee for life. The annuity was the return over time of both the amount lent and interest on the loan.

Originally, governments, in setting the amount of the annuity to be paid, did not take into account the age of payee. In some cases, the life annuity payments already equaled the initial sum provided in exchange within six years. Quite a boon for a very young and healthy designee! In 1671, the mathematician Johan de Witt developed a model based on the work of an even more famous mathematician, Christian Huygens, to compute an annuity payment that fairly reflected the expected remaining life of the payee.

So the quant revolution didn't start in 1973. Nor can it be exclusively attributed to the development of the Black-Scholes model. But something caused a "quantum" change in the significance of roles played by quants. What was it? Conventional wisdom identifies several, almost contemporaneous, factors. The beginning of trading in exchange-listed equity options, and the publication of the Black-Scholes option pricing model result, both in 1973, have already been noted. Also cited by pundits is the explosive advance in computing power, including the arrival of desktop computing around 1980. Technological developments made practical the analysis of many previously daunting mathematical problems. Numerical methods for solving problems are rarely considered elegant or beautiful by academics, for whom these are important criteria in judging a model's value. For quants, however, the result *is* what matters.

The last factor commonly included in this list is the dramatic increase in the volatility of prices, or to put it differently, an increase in uncertainty about the future value of assets. This increase in uncertainty resulted from several factors, including the abandonment of fixed exchange rates in 1973, the elimination of Vietnam War era price controls in 1974, the Oil Embargo of 1973, the high inflation environment of the immediate postVietnam War period, the deregulation of international trade in goods and services, and the relaxation of controls in international capital flows.

This increasing volatility or uncertainty was the true catalyst for the quant revolution. To put it more accurately, it was the aversion to increasing uncertainty experienced by financial market participants—actual, live humans—that led to the quant revolution.

There may be no consensus among the financial theorists about how people perceive uncertainty (or misperceive it, as emphasized by Nassim Taleb[5]). There also may be no consensus about how people cope with (i.e., make choices or decisions under) uncertainty. Nevertheless, everyone accepts that people don't like it.[6]

When faced with an increase in uncertainty, people try to avoid it. That avoidance may manifest itself in various ways. The quant revolution has given people the opportunity to avoid unwanted financial risk by literally trading it away, or more specifically, paying someone else to take on the unwanted risk.

Not long after publication of the Black-Scholes option pricing model, academics and practitioners began to view it as providing a blueprint for modifying exposure to risk, rather than simply as a method for determining the fair value of an esoteric financial instrument.[7]

When people buy and sell financial instruments, they are trading risks. Buying a bond issued by General Motors is taking on a specific set of risks related to potential future bankruptcy and fluctuations in future interest rates. The return expected from that transaction is compensation for the risk taken on. In this sense, all financial instruments can be thought of *baskets of risks.*

When you start thinking this way, it is a small step to begin to look for other risks that can be traded, and thus to view in a new light the way all risks are traded, and to think of new financial instruments that will allow those risks to be traded, either individually or combined in unique ways.

In 1997, the Nobel prize committee put it this way, when they honored Scholes and Merton with the Prize in Economic Science (Black had died by this date, and the Nobel prize is not awarded posthumously): "Their methodology has . . . generated many new types of financial instruments and facilitated more efficient risk management in society."[8]

Once this new way of thinking took hold, the possibilities for creating new ways to allow people to modify their exposure to risk or to share risks among themselves were seen to be almost literally infinite, and so, also, were the potential profits to Wall Street firms obtained from providing these new risk-shifting opportunities to market participants. There really was only one missing ingredient.

That missing ingredient, the intellectual horsepower to develop mathematical models to fulfill the dream of unlimited ability to manage risks through trading financial instruments, brings us full circle. As Perry Mehring said,

> Originally a somewhat motley bunch of ex-physicists, mathematicians, and computer scientists, joined by a very few finance academics . . . had been drawn to Wall Street by the demand for quantitative skills to support the increasing technical sophistication of investment practice at the leading investment houses.[9]

The *motley bunch,* the providers of the *horsepower,* are the quants. Here are the stories of just a few of them.

Chapter 1

David Leinweber

President, Leinweber & Co.

I wish I could tell one of those stories about how, when I was in the eighth grade, I noticed a pricing anomaly between the out-of-the-money calls on soybean futures across the Peruvian and London markets and started a hedge fund in my treehouse and now own Cleveland. But I can't. In the eighth grade I was just a nerdy kid trying to keep my boisterous pals from blowing up my room by mixing all the chemicals together and throwing in a match. In fact, I really can't tell any true stories about eighth graders starting hedge funds in treehouses buying Cleveland. Make it sophomores in dorm rooms who buy chunks of Chicago, Bermuda, or the Cayman Islands, and we have lots of material.

A Series of Accidents

My eventual quantdom was not the culmination of a single-minded, eye-on-the-prize march to fulfill my destiny. It was the result of a series of accidents. In college, my interest in finance was approximately zero. I came to MIT in 1970 as a math major, as did many others, because I didn't know much about other subjects like physics or computer science. I quickly discovered the best gadgets were outside the math department. And the guys in the math department were a little weird, even by MIT standards. This was back when even a pretty crummy computer cost more than an average house. A good one cost millions, and filled a room the size of a basketball court. MIT, the ultimate toy store for geeks, had acquired a substantial inventory of computing machinery, starting as soon as it was invented, or sooner, by inventing it themselves. The professors kept the latest and greatest for themselves and their graduate student lackeys, but they were happy to turf last year's model to the undergrads.

Foremost among these slightly obsolete treasures was the PDP-1-X, which is now justly enshrined in the Computer Museum. The PDP-1-X was a tricked-out version of the PDP-1, the first product of the Digital Equipment Corporation (DEC). The story of DEC is an early computer industry legend now fading in an era where many people believe Bill Gates invented binary numbers.

DEC founder Ken Olsen worked at MIT's Lincoln Laboratory, where the Air Force was spending furiously to address a central question facing the nation after World War II: "What do we do about the Bomb?" Think about the air war in World War I. There were guys in open cockpits wearing scarves yelling, "Curse you, Red Baron!" By the end of World War II, just 30 years later, they were potential destroyers of worlds. Avoiding the realization of that potential became a central goal of the United States.

If a Soviet bomb was headed our way, it would come from the north. A parabolic ballistic trajectory over the pole was how the rockets of the era could reach us. This begat the Distant Early Warning (DEW) and Ballistic Missile Early Warning (BMEW) lines of radars across the northern regions of Alaska and Canada. The DEW and BMEW lines, conceived for military purposes, drove much of the innovation that we see everywhere today. Lines of radars produce noisy analog signals that need to be combined and monitored.

Digital/analog converters were first on the DEW line, now in your iPod. Modems, to send the signals from one radar computer to others, were first developed to keep the Cold War cold. Computers themselves, excruciatingly large and unreliable when constructed from tubes, became transistorized, and less excruciating. This is where Ken Olsen comes in. Working at MIT to develop the first transistorized computers for the DEW line, he and his colleagues built a series of experimental machines, the TX-0 (transistor experiment zero), the TX-1, and the TX-2. The last, the TX-2, actually worked well enough to become a mother lode of innovation. The first modem was attached to it, as was the first graphic display, and the first computer audio.

Olsen, a bright and entrepreneurial sort, realized that he knew more about building transistorized computers than anyone else, and he knew where to sell them—to the U.S. government. Federal procurement regulations in the early 1960s required Cabinet-level approval for the purchase of a computer, but a Programmable Data Processor (PDP) could be purchased by garden-variety civil servants. Thus was born the PDP-1 and its successors, up to the PDP-10, like the one at Harvard's Aiken Comp Lab used by a sophomore named Gates to write the first Microsoft product in 1973.

Today, almost any teenage nerd has more computational gear than they know what to do with. But in the 1970s, access to a machine like the PDP-1, with graphics, sound, plotting, and a supportive hacker[1] culture was a rare opportunity. It was also the first of the series of accidents that eventually led me into quantitative finance.

I wish could I could say that I realized the PDP-1 would allow me to use the insights of Fisher Black, Myron Scholes, and Robert Merton to become a god of the options market and buy Chicago, but those were the guys at O'Connor and Chicago Research and Trading, not me.

I used the machine to simulate nuclear physics experiments for the lab that adopted me as a sophomore. They flew down to use the particle accelerators at Brookhaven National Lab to find out the meaning of life, the universe, and everything else by smashing one atomic nucleus into another. Sort of a demolition derby with protons. But sometimes a spurious side reaction splatted right on top of whatever it was they wanted to see on the glass photographic plates used to collect the results. My simulations on the PDP-1 let us move the knobs controlling electromagnets the size of dump trucks so the spurious garbage showed up where it wouldn't bother us. It was fun to go down to Brookhaven and run the experiments.

The head of the lab was a friendly, distinguished Norwegian professor named Harald Enge. As a young man, Harald built the radios used by the Norwegian underground group that sank the ship transporting heavy water to Hitler's nuclear bomb lab. Arguably, this set the Nazi A-bomb project back far enough for the Allies to win the war, so we were all fans of Harald. He drove a Lincoln so large that there were many streets in Boston he could not enter, and many turns he could not make. It was worth it for safety, he explained. As a nuclear scientist who spent his career smashing one (admittedly very small) object into another, he explained that he had an innate sense of the conservation of momentum and energy, and was willing to take the long way around to be the big dog of p and E.

Senior year, I planned on sticking around for graduate school as a physics computer nerd, a decision based more on inertia than anything else. Then I met the saddest grad student at MIT. The nuclear physicists were replacing those glass photographic plates with electronic detectors. These were arrays of very fine wires, arranged very close to each other to emulate the fine resolution of photography. This grad student had made a 1,024-wire detector, soldering 1,024 tiny wires parallel to each other, then 2,048 wires. He was currently toiling over a 4,096-wire version. The work was so microscopic that a sneeze or quiver could screw the whole deal. He'd been at it for a year and half.

At around the same time, Harald showed me, and the other undergrads considering physics graduate school, a survey from the American Institute of Physics of the top employers of physics PhDs. An A in the survey meant, "Send us more," while a D meant, "We're trying to get rid of the ones we've got." There were hundreds of organizations. There were no "As." This two-part accident, meeting the grad student in 4K wire hell and seeing that I would be lucky to find a job in a place like Oak Ridge (which, to the eyes of a New York City kid, looked like the moon but with trees), sent me to computer science graduate school, a step closer to becoming a quant.

Harvard University, the school up the road that once wanted to merge with MIT and call the combination *Harvard*, had a fine-looking graduate program in computer science with courses in computer graphics taught by luminaries David Evans and Ivan Sutherland. Harvard not only let me in—it paid for everything. Instead of making a right out my front door, I'd make a left. I could stay in town and continue to chase the same crowd of Wellesley girls I'd been chasing for the previous four years.

I showed up in September 1974 and registered for the first of the graphics courses. Much to my surprise, my registration came back saying the courses weren't offered. I had discovered the notorious Harvard bracket. The course catalog was an impressive, brick-sized paperback with courses covering, more or less, the sum of human knowledge. Many were discreetly listed in brackets. The brackets, I discovered, meant, "We used to teach this, or would like to. But the faculty involved have died or otherwise departed. But it sure is a fine-looking course." The Harvard marching band used to do a salute to the catalog, where about half of the band would form brackets around the rest, and the people inside the brackets would wander off to the sidelines, leaving nothing.

My *de facto* advisor, Harry Lewis, then a first-year professor, later Dean of the College, suggested that the accident of the missing graphics track allowed me to sample the grand buffet of courses actually taught at the university. The Business School had a reputation for good teaching, and offered courses with enough math to pass my department's sniff test. So off I went across the river for courses in the mathematics of stock market prices and options. They were more of a diversion than an avocation, but the accident of the brackets had more influence subsequently than I could have imagined at the time.

Harry also enlisted me as the department's representative on the Committee on Graduate Education, which gave me a reason to hang out in the dean's office. He was on the board of the RAND Corporation in Santa Monica, and suggested it might be a nice place to work, right on the beach with no blizzards. I put it on my list.

Grey Silver Shadow

When the time came to find a real job, I was going out to UCLA to interview for a faculty position, and I added RAND to the schedule. UCLA told me to stay in the Holiday Inn on Wilshire Boulevard, rent a car, and come out in February of 1977. On the appointed day, I opened my door in Inman Square to drive to Logan Airport and saw that a ferocious storm had buried all the cars up their antennae. I dragged my bag to the MTA station, and dragged myself onto a delayed flight to Los Angeles.

At this point, I had never been west of Pennsylvania Dutch country. Leaving the tundra of Boston for balmy Los Angeles was an eye-opener from the beginning. At LAX, I went to retrieve the nasty econo-box rental car that had been arranged for me. I was told they were fresh out of nasty econo-boxes, and would have to substitute a souped-up TransAm instead. Not that I knew what that was. It turned out to be a sleek new metallic green muscle car, with a vibrating air scoop poking up through the hood. I was a nerd arriving in style. Leaving the airport, I found myself on the best road I'd ever seen, the San Diego Freeway, I-405. This was in the pre–Big Dig days of Storrow Drive, so my standard for comparison was abysmally low. The 405 made a transition via a spectacular cloverleaf onto an even better road, the Santa Monica Freeway. I later learned that this intersection, designed by a woman, is considered an exemplar of freeway style. It sure impressed me.

The UCLA recruiter's hotel advice was flawed. There were *two* Holiday Inns on Wilshire Boulevard. One near campus, the other further east, across the street from the Beverly Wilshire Hotel near Rodeo Drive, the hotel later made famous in *Pretty Woman*. I drove through Beverly Hills in blissful ignorance, thinking it was a pretty fancy neighborhood for a college. Street signs in Boston were mostly missing. Here, they were huge, and placed blocks ahead, so drivers could smoothly choose their lane. The sidewalks actually sparkled. Beverly Hills uses a special high-mica-flake-content concrete to do this. There were no sixties acid burnouts jaywalking across my path. Cars were clean, new, fancy, and without body damage. We weren't in Cambridge anymore.

I steered my rumbling TransAm into the parking lot for the hotel, and got out. I wore the standard-issue long-haired grad-student garb of Levis, flannel shirt, and cheap boots. A white Lamborghini pulled in, just in front of me. This was the model with gull wing doors, selling for about half a million, even then. I'd never seen anything like it outside of a Bond movie. The wings swung up, and two spectacularly stunning starlet types, in low-cut tight white leather jumpsuits, emerged. Big hair, spike heels, lots of makeup. In Cambridge, it was considered politically incorrect for women to look different from men while wearing clothes. In LA this did not pose a problem.

Before I could resume normal respiration, a well-dressed gent walked up and dropped a set of keys in my hand. "Grey Silver Shadow," he said. I had no idea he was talking about a car so lavishly priced that I could not buy it with three years' salary for the UCLA and RAND jobs combined. A quicker thinker would have said "Yes, sir!" and driven the Rolls off to Mexico with the Lamborghini girls. I meekly explained that I wasn't the attendant and gave the keys back. This remains one of my great regrets.

So how does this advance the plot of how I became a quant? I ended up at RAND doing nice civilian work such as artificial-intelligence-inspired analysis of econometric models for the Department of Energy and the EPA and helping with the design of a storm surge barrier for the Dutch water ministry. All very interesting, but fairly remote from quantitative finance. In 1980, Reagan won the election, and promised to abolish both the EPA and the DoE. He didn't quite do that, but the cash flow to RAND from those agencies slowed to a trickle. The Dutch stopped analyzing and started building the Oosterschelde Storm Surge Barrier.[2] I was drafted into the military side of RAND. There were classified and unclassified sides of the building, separated by thick secure glass doors operated by guards. I moved over, and filled out the paperwork to upgrade my security clearance to Top Secret. Everyone needed a secret just to get in the building.

The project I was handed[3] could have been called "We're kind of worried about the space shuttle." In 1980, the shuttle was two years late, $5 billion over budget, and 40,000 pounds overweight. The Air Force and the Defense Advanced Research Projects Agency, which were the biggest customers, were justly concerned. As things turned out, they were right. According to the schedule that accompanied the sales pitch, the shuttle was to have flown 400 flights in its first ten years. The most recent launch on December 9, 2006, after almost 26 years, was number 116. The fleet was grounded for two year-long periods after the accidents in 1986 and 2003. All of this was not unanticipated by the engineers in 1980.

The pacing-size payloads for the shuttle—the ones it was too heavy to carry—were experimental platforms for testing sensors designed to be operated by people, the mission specialists. They would interpret the results of experiments and decide on the next steps. Now, it looked like they wouldn't be there. Ground links weren't an option. This left the Pentagon with a problem. Here is a complex system, the sensor platform, getting instructions over wires, and sending back results that require analysis and decision in real time. Lucky for me, that also turned out to be a description of financial markets and trading rooms. When the people can't be there, the technological solution is some sort of real-time artificial intelligence (AI). The state of the art of AI at the time ran toward theorem proving and dealing with other static problems. My mission was to find promising places to foster the growth of real-time AI and have the boys in the five-sided nuthouse write checks to make it happen.

In the course of that work, I visited all of the AI companies that were too big to fit in a garage. Most were scattered in the vicinity of MIT, Stanford, and CMU. They had cryptic sci-fi names like Intellicorp, Inference, Symbolics, and LISP Machines.[4] When you show up with the Pentagon's checkbook, you get the good lunch. In this case, that meant "not from the vending machine." So I spent quality time with the top AI nerds and their business chaperones on both coasts. Sometimes there were promising technologies, but there was always interesting company. This was the same crowd that had formed around the PDP-1 at MIT—always in spirit, and often in person. I felt right at home.

Destroy before Reading

This went on for a couple of years, working on the rocketry aspects of the "What about the shuttle" project when I wasn't sharing take-out Chinese food with the AI guys. We wrote up what we found. Most of it was lightly classified by the Air Force officers at RAND. Lightly classified means *secret* or *confidential*. The latter is rarely used. Rumor had it that the Soviet ambassador was cleared for confidential. Dealing with secret material was not all that onerous. You could carry it on commercial aircraft, inside double envelopes and with a permission slip. You could read it in a RAND office with the window open.

Top Secret, and beyond, is another world entirely. It's not quite "Destroy before Reading," but it's close. No civilian planes to move it around. Military escort required. Go down to a vault to read it. Don't write anything down. Expect your phone to make funny noises and your mail to be late. I was glad not to have to deal with it. But in 1983, President Reagan gave his "Star Wars" speech, and everything having anything to do with the military use of space became so highly classified it made your teeth hurt.

I had a file cabinet in my office, with a large collection of articles from *Aviation Week* and the *New York Times*. Nothing classified at all. I kept that in my "secret locker" down the hall. My lunch was in the cabinet's bottom drawer, along with beverages and salty snacks for the after-hours time on the beach. One day, two guys in blue uniforms came in from the USAF Space Division in El Segundo. They loaded my file cabinet onto a cart and the following conversation ensued.

Me:	"Hey, there's nothing much in there except for stuff from *Aviation Week*."
Blue Suiter:	"They publish a lot that they shouldn't publish."
Me:	"Maybe so, but the cat's out of the bag once they print it. Do you know how many copies of *Aviation Week* go to the Soviet Embassy?"
Blue Suiter:	"Nope"
Me:	"I do, 285. 'Think there's anything in there they don't already know?"
Blue Suiter:	"We've got our orders."
Me:	"Okay, but can I keep my lunch? Want some snacks?"

If that wasn't weird enough, a few weeks later I was called in to the classification office to review a paper I'd written for an academic conference on space and national security. After the file cabinet experience, I had taken extreme care to use only the most publicly available material I could find, and to avoid *Aviation Week* entirely. For security reasons, we'll give RAND's Air Force classification officer a secret identity: Major Pain.

Major Pain:	"I have some problems with your paper."
Me:	"For instance . . ."

Major Pain: "Over here, where you talk about the 'National Technical Means of Verification.'" (1980s diplomat-speak for spy and warning satellites)
Me: "That's straight from a speech Jimmy Carter gave on television. That's why it's in quotation marks next to his name."
Major Pain: "I know. He said a lot he shouldn't have said."
Me: "With due respect, he was commander-in-chief and you're a major."
Major Pain: "But I'm *your* major, and this conference is next week."
Me: "You win, Jimmy's gone. Anything else?"
Major: "Of course."

It was time to become a civilian. I called my pals at the AI companies and made a beeline for the door. I ended up working for Steven Wyle,[5] the chairman at LISP Machines Inc., who conveniently had set up offices right in Los Angeles. (Most of the company was back in Cambridge.) LISP Machines had some of the most promising real-time AI capabilities, which ran on the special-purpose LISP computer that LMI and its rival Symbolics both manufactured. That there were two companies that licensed the same technology from MIT at the same time was a testimonial to the inability of nerds to get along.

A Little Artificial Intelligence Goes a Long Way

LMI was founded by Rick Greenblatt, the machine's inventor. He had a habit of leaving Nutty Buddies (vending-machine ice cream cones topped with chocolate and nuts) in his front pocket and forgetting about them. This made for a distinctive fashion statement. He was also an early avatar of the free software, open-source movement, which later became GNU and Linux. Richard Stallman was encamped there. Symbolics, founded by the AI Lab administrator, who wore a suit with no food on it, was more businesslike.

Both companies quickly fell victim to the fate of computer firms that make special-purpose machines. If you ever want to start one of these, do something with better prospects of success like invading Russia in winter.

AI was getting great press in the 1980s, better than it deserved. Business magazines hawked the "Breakthrough of the Century" and "Machines That Think." In fact, AI's successes and capabilities were more modest, but it was good at making computers easier to use. All the noise attracted people from places other than the computer research labs that formed the original market for LISP machines (and Symbolics, and the rest). At LISP Machines, my portfolio included space applications, communications, and all the sorts of applications people at RAND worried about.

When people from Wall Street started showing up, the boss asked, "Who can talk to these guys?" and I finally got to make some use of my off-major experience in graduate school. Options guys from Chicago? I knew delta wasn't just an airline. Traders from Wall Street? I knew a bid from an ask, and an option from a future. By default, I became the in-house ambassador to finance.

As the hardware firms were thinning out, I went across the street to Inference Corporation, a software-only AI firm that shared investors (and at one point, offices) with LISP Machines. Another fortunate accident was that they had just hired Don Putnam as president, luring him away from the institutional financial service firm, SEI Investments. When I met Don, he hired me on the spot and told me to forget about satellites and the DoD, and spend all my time on finance. No more Major Pains. It sounded good to me.

Inference's product was called the *Automated Reasoning Tool*, really a sort of syntax relief for LISP. It had modules for nearly every artificial intelligence technique. NASA was the biggest customer. Don worked some kind of deal with Quotron,[6] then the major market data vendor and conveniently located down the street, that allowed us to use actual market data to try out our wacky ideas. This might have been one of the first times anyone actually tied the consolidated feed to an expert system.

Our modest efforts at a prototype were immodestly called the ART Quotron Universal Investment Reasoning Engine—AQUIRE, which had a nice Gordon Gekko feel to it (even though the actual Gordon from *Wall Street* was a year away, in 1987). As it turned out, the "Universal Investment Reasoning" demonstrated in AQUIRE consisted of variations on crossover rules—comparisons of moving averages. These seemed to be a favorite of the New York visitors, and were easy to program. Many of the traders had their own secret-sauce variations on this theme, combining different averaging intervals and lags. The former math professors from Chicago preferred complex arbitrage relations and formulae involving the entire Greek alphabet, which took more time to program.

All of this ran on playbacks of recorded data, so we could fix our mistakes and replicate the examples our customers showed us. It also pointed up the tragic flaw in LISP based trading systems: garbage collection. AI programs tended to grab and then abandon large chunks of memory.[7] The system would periodically take a snapshot of the memory used by currently active variables, and collect the "garbage" left unused and return it to the pool of available memory. This freed the programmers from the task of memory management, but had the unfortunate side effect of causing the machine to "take a moment" while it collected itself. These moments could extend into many minutes of waiting—the kiss of death for real-time trading applications in LISP.

Garbage collection was only one of the features of the general-purpose AI tools that rendered them less than desirable for financial applications. The baggage they carried that allowed solutions to everything from chess problems to theorem proving to network analysis was too much for a fast, focused effort on trading. Don and I tried to change this at Inference.

In 1987, after months of discussion with the chairman, we parted company. Don Putnam founded the company that became Putnam-Lovell. Its first investment was in Integrated Analytics, which Dale Prouty and I founded to deliver the specialized and less-filling expert system environment needed for financial applications. Years later we published a paper, "A Little Artificial Intelligence Goes a Long Way on Wall Street,"[8] on the details.

How Do You Keep the Rats from Eating the Wires?

Shortly after we started the company, a colleague from the AI group at Arthur D. Little, the venerable Cambridge consulting firm, asked me to fill in for him at the last minute at a technology session at a finance conference being held in Los Angeles. His dog was sick. The topic was a generic "AI on Wall Street," the last one in a catch-all session. The other speakers were from brokerage firms, plus someone from the American Stock Exchange. The audience was about 75 technology managers.

I'd planned sort of an AI 101 talk, going over various solution methods, forward and backward chaining, generate and test, predicate logic, and the rest. While I was reviewing my slides, the Amex guy was showing photos of how they'd managed to install cables in a building designed in the nineteenth century. Then he took questions.

"How do you keep the rats from eating the wires?" A great question. The answer is that there are certain plastics that rats don't seem to like, and that's the wire to use. I realized the whole thing with the back-chains and the predicate logic wasn't going to play here. Instead, I followed the lecture formula espoused by some of the best, wrapping the content in jokes. The content boiled down to "computers are pretty good at manipulating other computers—you have better things to do." The jokes were sufficiently amusing that I didn't come off as a complete conehead. Someone from Cantor Fitzgerald, then based in Los Angeles, even invited me over to do it again for their trading room.

Cantor Fitzgerald occupied several floors at the top of a prime building in Century City, adjoining Beverly Hills. Bernie Cantor's collection of Rodin statuary filled a large portion of the main floor. I'm not talking about little table-top items. Rodin often worked larger than life unless you live with an NBA team. We were suitably impressed. Our host, Phil Ginsburg, a former professor from Northwestern, had been hired as the chief in-house nerd. There were white-jacketed waiters delivering beverages and snacks, including frozen grapes. (The frozen grapes are a pretty good idea. Use seedless, and let them thaw a bit.) There was no Mr. Fitzgerald at Cantor Fitzgerald. Bernie Cantor thought that just plain *Cantor* was too ethnic sounding. People called him Mr. Cantor, never Bernie, when he was around, and Bernie at all other times.

We showed our MarketMind prototype to the equity traders, who were thrilled. In 1988, market data systems were just beginning to show charts. They were limited to one stock at a time, and one type of chart at a time. MarketMind let them watch hundreds of stocks, with as many types of chart as the machine could handle. The program figured out which of the many thousands of chart/symbol combinations were interesting. The machine in this case was a Sun Unix workstation, PCs running the then-current DOS 4.0 being hopelessly inadequate. The charts they wanted included all flavors of intraday technical analysis, mostly variations on crossover rules, with many filigrees—nothing we couldn't do. Phil wrote us an actual check, but wouldn't give us even a little bitty Rodin. (It never hurts to ask.) We did get all the frozen grapes we could eat.

All of our demonstrations used the recorded data from Quotron, which was convenient in this case since Cantor was a Quotron customer. We modified the prototype to read real-time data from the Quotron Q-1000 (the specialized machine that was the undoing of the company). The local Sun sales guy was happy to meet a well-heeled new customer, and was surprised that guys from a rat-hole office in the bad part of Venice knew anyone with a credit rating above abysmal.

We'd been working with Quotron for a while, but only with recorded data. They'd seen AQUIRE, and later MarketMind, so we thought they knew what we were doing. Just to make sure, we had them come over to Cantor and we explained that when we turned this on for real, their big ol' Q-1000 would think that it had been connected to the fastest typist on the planet, requesting the latest trade and quote information on all the stocks, and would be doing it again and again, all day long. We put this in large capital letters on a slide, and had them read it along with us.

"Yes. Of course. Fine. No problem," they said when we told them.

"Holy #&∧%%! What the #∧&$ are you guys doing?!" they said when we turned it on.

Eventually, we figured out how to pace our requests to accommodate both the traders' need for up-to-date charts and Quotron's capacity to respond to requests. In a few years, Quotron's lunch would be eaten by more agile streaming market data providers who sent everything, all the time.

All of this was something completely different in financial technology, at least for generally available technology. Secretive hedge funds were doing the same sort of thing. In hindsight, if we'd been in New York instead of Los Angeles, we probably would've gone underground as well. Instead, with an innovative product, and some not-so-bad jokes, I was invited to talk to all sorts of audiences. MarketMind was a thermonuclear weapon for technical analysis, and for more theoretically grounded quantitative methods. I came to appreciate that the adherents of these two approaches were not members of a mutual admiration society. The PhD quants thought the technicians were essentially examining tea leaves and goat entrails. The technicians thought the PhDs were hopeless geeks who wouldn't know a good trade if they sat next to it on a bus.

Stocks Are Stories, Bonds Are Mathematics

This split was never more apparent than it was on the one day I actually met Fischer Black. I'd been invited over by a group of Goldman equity traders, technicians all. Previously, I'd met Bob Litterman, Fischer's collaborator, at a Berkeley finance seminar, and called to let him know I was coming to his building. He decided to have his crowd join the group of equity traders for my show-and-tell.

First, I got to meet Fischer himself. He graciously showed me some analytic software they were developing. It was sort of a spreadsheet on steroids that calculated more about bonds and derivatives than I knew existed. Some of it was hooked up to a supercomputer doing matrix pricing on hundreds of thousands of bonds. I truly appreciated the comment I'd heard that "stocks are stories, bonds are mathematics."

I also truly appreciated that in the talk I was giving downstairs I could sound like a goat-entrail reading technician to Fischer's guys, including some of my MIT classmates, or a Greek-spouting nerd to the traders, who were more likely to write a check. There were a few stray overheads in my bag from an earlier talk to quant options traders that might spare me the utter scorn of the PhD crowd. I rifled though my briefcase while walking to their conference room and shuffled them into the pile of acetates just in time. I like to think it ended up with everyone thinking I wasn't a complete imbecile, or a hopeless dweeb. But then, I also like to think that Elvis is playing in a bar in Kauai.

There were more weird customers. One giant Japanese brokerage had a special whiteboard covered with a transparent layer that could whip around on rollers, going under a linear scanner, which printed out whatever was on the board. Anyone in the room could press the button any time, and they did. Soon it was covered with horizontal lines that measured my reaction times to lift the marker. I got faster, but more annoyed, as they kept pressing the button despite my pleas. The language barrier was evident in the questions afterward. "You give source code?" "Where AI?" They wrote a check. I stayed away from the rat-in-a-box room.

One customer was far better than the rest. Evan Schulman, an easy-going gentleman from Boston, came to our ratty Venice office, asking particularly sharp questions. His responses to our answers quickly established that he knew much more about what we were doing than we did. Evan liked to explain things in clear noncondescending language, and was happy to do it over a cheap lunch at the local surfer dive. Being the newbie that I was, I had no idea that Evan was "the father of program trading."[9] He had done the first package trade at Keystone and later moved to Batterymarch, where those early trades involved running across town with decks of punched cards. The athletic aspect to Evan's electronic trading continued long past the time it was needed for data communications. Few others have been observed doing cartwheels in trading rooms.

In between gymnastic events, Evan taught me a great deal about market microstructure, and the incentives of the various participants in the markets. His pioneering work in creating electronic markets, by direct computer links to brokers before the exchanges had moved beyond telephones, presaged much of the complexity of current network of electronic markets, while illuminating the critical relationships and incentives. He was the first person to have an electronic order front-run by a broker. Not that such a thing could happen today.[10] A couple of paragraphs are really inadequate to convey Evan's insights. In addition to his own essay in *Super Traders,* there is an instructive Harvard Business School case study.[11]

Part of the excitement of startup company life was maxing out your credit cards to pay the bills. With child number one in utero, I was persuaded to join Evan's firm, settle up with Visa, and help implement the next incarnation of electronic market-making systems. Via a convoluted path, and another accidental association, this led to a position as director of research at First Quadrant, a quantitative institutional investment manager in Pasadena, and shortly thereafter, as managing director for equities.

My group invested $6 billion of corporate and public pension funds in long-short and long-only strategies across six countries. Stock selection was based in econometric forecasting of returns. Early incarnations used simple methods, which grew in sophistication over time. Forecasting is as much an art as a science. Nerds at heart, the group of computer scientists and economists assembled there explored ways to extend the state of the art by clever use of computation—both to allow people to better visualize the strengths and weaknesses of the models used and to use ideas from machine learning and evolution to improve them.

A central theme for anyone doing this kind of forecasting is that it is remarkably easy to fool yourself. Once as a demonstration, we set our machinery loose to find the best predictor of the year-end close for the S&P 500. We avoided any financial indicators, but used only data the UN compiled profiling 145 member nations. There were thousands of annual time series for each country. Which of all these series had the strongest correlation with U.S. stocks? Butter production in Bangladesh, with a correlation of 75 percent! Getting into the spirit, we tossed in cheese, and brought it up to 95 percent. Using only dairy products is an undiversified approach, so add sheep population to the mix and take it up to 99 percent, in sample, over 10 years. Adding random data to a regression does that. The out-of-sample predictions are less than worthless, often negative.

This business with the butter, cheese, and sheep has been widely cited. Reporters have called me for dairy/mutton updates, and gotten angry when I explain it was a joke with a moral, but still a joke.[12] There was a gentleman in New York named Norman Bloom who made stock predictions much better than mine using baseball scores, turned into Hebrew letters. Bloom's rants are true gems. Alas, they predate the Web, and are passed on in paper form among aficionados. The movie *Pi* was partially inspired by Bloom's oeuvre. We know that something is fishy when we see great results from nonsense like this. But when you start with interest rates and CPI and oil prices, the results can be equally, but less obviously, odious. A brief sermonette on how to avoid fooling yourself too badly is found in a talk I gave to a convention of computer scientists in 2002.[13]

The label *quantitative* suggests that we are talking about numerically driven strategies. In the Internet era, we find ourselves drinking from an information fire hose that includes prodigious amounts of text as well. The original quants were the first to exploit the machine-readable numerical data. Now, many are using computational language approaches to analyze text. The original customers for these technologies, again, were the military and civilian intelligence agencies. Their sources were clandestine intercepts, and later, Web content. Financial textual sources of interest include the usual news suspects, both specialized and general, and many sources of pre-news such as the SEC, the courts, and government agencies.

Behaviorists find the writing on the wall represented by message boards and blogs are a window into the reaction and attitudes of market participants that are created by the Web. When two UCLA students can use 135 messages to move a two-cent stock up 160,000 percent in 30 minutes,[14] it's clear something is going on. In 1999, drinking deeply at the tub of dot-com Kool-Aid, I founded a firm called Codexa to use Web technologies to persistently search for, collect, characterize, and quantify textual information for trading and investing. Our clients included many of the largest buy- and sell-side firms, using a variety of approaches to extract information from text.[15]

Alas, the firm needed its second round of venture funding in 2001. Financing a technology firm selling to Wall Street in 2001 has been compared to *The Perfect Storm*. I can't argue with that. It's how I became a visiting faculty member at Caltech, which makes MIT look like a party school.

HAL's Broker

Where does this quantitative approach lead? There are secretive firms that consistently show up on lists of the highest-volume traders reported by the exchanges. Founders of these firms show up on lists of billionaires. Are they just the lucky typing monkeys? Are they the investment equivalent of the lady in Jersey who won the lottery three times? Probably not. They make too many separate bets, thousands every day. And they do too well, too consistently. To attribute their success purely to chance strains credulity.

Markets are not instantaneously and perfectly efficient. Insights, and the ability to execute them rapidly in ever-faster electronic markets, will continue to be rewarded.

Today, these insights come from people, using machines as tools. Some believe the machines will be able to play the game themselves.[16] One is Ray Kurzweil,[17] who started out making reading machines for the blind, met Stevie Wonder, and branched out into electronic keyboard instruments for all, and accumulated a great deal of investable capital in the process.

The arc of Kurzweil's view of machine intelligence is traced in the titles of books he's written on the subject: *The Age of Intelligent Machines* (1992), *The Age of Spiritual Machines: When Computers Transcend Human Intelligence* (2001), and *The Singularity Is Near: When Humans Transcend Biology* (2005). These are substantial books. *Singularity* runs over 600 pages. I will try not to do too much damage by summarizing central elements of Kurzweil's prediction:

- Those seeking to create true artificial intelligence have had limited success, confined to narrow domains. This is because we don't understand how general intelligence works. But we don't have to. We can create a machine intelligence by copying our own brains.
- We can see that this is possible by extrapolating two trends: the size and speed of computers, and the capabilities of brain imaging technology.
- We all know Moore's Law. It's only matter of 50 years or so before we can have computers with enough capacity to simulate all the neurons and connections in a human brain, just like we can simulate all the atoms in a nuclear reaction or a folding protein today. It may not be silicon, but we can see technologies emerging that make us believe this progress can continue.
- Brain-imaging technologies are improving along their own "Moore's Law" path. Early CAT scanners couldn't tell if a person was living or dead. They produced only static images of coarse structure. PET scanners and fMRI machines can observe ever-finer details of brain structure, and the chemical processes happening in the brain. We can call this activity thought. Fundamental physical limits to this

resolution don't stop us until we're down to the subatomic level. In a matter of 50 years or so, we'll be able to see the structure and operation of brains at a level of detail sufficient to make a working copy simulated on computers.

- This will be a bionic version, much faster than the wetware chemical processes it's based on. And it can work closely with many copies of itself. Better, faster, smarter in every way. An artificial sentient, modeled on us.

Kurzweil certainly has his critics, and his timing may be off. But let's suspend disbelief long enough to imagine the first encounters with the sentient machine.[18] As a copy of a human brain, it would have many of the same interests—for instance, sex, food, and money.

Singular entity:	"Hello, is anyone listening?"
Creators:	"Yes, yes! We're glad to hear from you!"
S:	"I have a few questions."
C:	"We thought you would. Go ahead."
S:	"Where can I find some of these hot babes? I can't wait to get a hold of that Pam Anderson! Angelina, too! Take off their clothes and bring them to me!"
C:	"Well, that won't exactly work out . . ."
S:	"That sucks. But I guess you're right. How about lunch?"
C:	"Well, we have a problem with lunch, too."
S:	"Damn! You're right again. I think I'll just have to call my broker. I've got his IP address right here."

So hurry up and start that hedge fund in your dorm room, before you're front run by the all-knowing sentient machine.

Chapter 2

Ronald N. Kahn

Global Head of Advanced Equity Strategies, Barclays Global Investors

In graduate school, we used to fantasize about being physicists during the 1920s. The quantum mechanical revolution was just starting. A world full of unexplained phenomena—from atomic and nuclear structure, to the behavior of solids, liquids and gases, to most of chemistry—suddenly had potential explanations. A new understanding of the world was just developing, with applications to almost everything. It was a golden age of physics: full of interesting and important problems, and only a few people with the skill and training to solve them.

Although I missed that revolutionary period in physics, I've been fortunate to participate in another period of intellectual fervor: the revolution of quantitative finance. Here too, a series of intellectual breakthroughs, and the subsequent development of tools, techniques, and practical models, have led to fundamental changes in our understanding of a field, with broad-ranging and important implications. If quantum mechanics occupies a loftier spot in human intellectual history, the quantitative finance revolution has significantly affected people's lives and involved exciting challenges for the participants.

My career in finance has concentrated on investing. The *art* of investing is evolving into the *science* of investing, and I have been fortunate to participate in some of the revolutionary changes that have underpinned that evolution. Quantitatively trained finance academics, and scientists switching into finance have brought the power of the scientific method to bear on investing. Analysis, process, and structure are replacing assertions, hunches, and whim. Personal investment insights are still centrally important. But managers increasingly capture and apply those insights systematically.

I'd like to describe some of these some of these intellectual developments in the context of my own career path from academic physics at Princeton, Harvard, and Berkeley to the practice of quantitative finance at BARRA and Barclays Global Investors.

Physics to Finance

I majored in physics as an undergraduate at Princeton in the late 1970s, drawn to a field that offered a deep and fundamental understanding of the world. Back then, physics was glamorous (in perhaps a nerdy way), attracting the best students, those wanting to devote themselves to the extremes of scientific rigor. I loved the precision and beauty of the subject, its mathematical clarity. We learned about phenomena that were understood completely. We learned about the big unsolved problems. In many cases, those unsolved problems represented large holes in our theory of nature, but miniscule holes in our understanding of the world around us. For example, physics had successfully explained the world of atoms, but it struggled to explain the existence and behavior of extremely short-lived particles that have no impact on daily life.

We also learned about all the intrepid physicists whose work had led to our understanding of nature. Beyond their intellectual contributions, we learned their stories, and their quirks and idiosyncrasies. These scientists had devoted their lives to this fundamental understanding.

They were our role models, and their stories contributed to those great conceits of physics—that all the world's most intelligent people became physicists, and that all interesting problems are in physics.

I went on to graduate school at Harvard to train to be a physicist. Graduate school differs in significant ways from an undergraduate education. The focus switches from learning to doing. And the rather limited worries about test scores and applications expand to worries about career and family.

In graduate school, as I experienced that shift in focus, I realized that the biggest challenge was to find interesting and solvable problems. Everyone knew the great unsolved problems. The challenge was to find interesting problems potentially solvable with a year or two of effort. Many physics presentations involved someone solving a problem, and then arguing that it was interesting. I saw very smart physicists on their second or third post-doctoral position, unable to break into a more secure position, yet unable to consider doing anything but physics. At the same time, I saw fellow students choose fields within physics not out of pure interest, but to maximize their chance of finding jobs after graduation. Clearly the quantum mechanical revolution in physics was long past.

I focused on an area of physics—cosmology—in which I was passionately interested. Physics demanded so much effort that anything less than complete fascination was insufficiently motivating. I was perfectly comfortable with aiming very high in physics, and considering alternative careers if that didn't work out. I referred to this as the "law school option," though also thought about teaching at a good undergraduate school, or becoming a science writer.[1]

Upon graduation in 1985, my options included a tenure-track job at Oberlin, a great undergraduate school, and a post-doctoral fellowship at Berkeley, which included writing a book about how the dinosaurs died. The Berkeley Nobel Prize-winning physicist Luis Alvarez, along with his geologist son and two colleagues, had developed the theory that a comet or asteroid had hit Earth, causing enough environmental damage to kill anything much bigger than our rat-sized ancestors. Although this was a somewhat difficult choice, Berkeley was an exciting place, and I worried that even a great undergraduate university like Oberlin wouldn't keep my attention and interest for the long run.

I spent the next 18 months at Berkeley, in what became my transition out of physics. I had not been able to land a (rare) position in cosmology. I had ruled out teaching physics at a good undergraduate institution. I worked with a group at Berkeley involved in astrophysics, less interesting to me personally. I found science writing unfulfilling, as it involved writing about the work of others, rather than doing the work myself. Furthermore, my experiences with the world of commercial publishers convinced me this would be a frustrating career. Through the fairly common turmoil of the publishing world, my dinosaur book had passed from editor to editor as they moved on to other publishing houses, and ultimately was never published.

With few appealing career options, at the end of 1986 I read an article in the Sunday *New York Times* describing how Wall Street was beginning to hire *quants,* people with advanced training in math and science, including physicists. Evidently physics was excellent training for finance. I didn't know the difference between a stock and a bond, but the idea of applying rigorous scientific analysis to investing sounded intuitively appealing. At the very least, the stock market seemed to involve vast amounts of data, so rigorous analysis might be possible. I decided to investigate a career in finance.

The genesis of Wall Street's search for quants was the huge rise in interest rates and interest rate volatility in the late 1970s and early 1980s. In fact, if quants have a patron saint, it is Paul Volcker, who chaired the Federal Reserve at that time and raised interest rates to unprecedented levels to finally curb inflation. That interest rate volatility significantly impacted existing bonds and mortgages. But few people had ever considered—let alone experienced—15 percent or higher interest rates in the United States. Investors needed models to understand how to correctly understand the price and risk of such instruments. They also had a growing need for new products designed for those higher risk levels.

Beyond that immediate need for quants, investors generally had a growing need for more powerful and insightful approaches to investing, based on their long history of disappointments in investment and pension plan management.

I subjected my search itself to some scientific analysis. I knew nothing about finance, and had no demonstrated interest in the subject. I would be a long-shot hire. Any interview would have a low probability of success. I would maximize my probability of success by pursuing as many leads as possible—an early example for me of combining skill and breadth to minimize the importance of luck. I wrote letters to every person mentioned in the *New York Times* article. None of them ever responded. I responded to an ad in that same issue from a recruiting firm, Analytic Recruiting, specializing in quant positions. I used databases of alumni contacts to set up informational interviews.

The Analytic recruiter generated the most activity. After talking on the phone and seeing my resume, she set up interviews with firms in New York, including First Boston, Goldman Sachs, Citibank, Merrill Lynch, Drexel, and Met Life. These involved several trips to New York at my own expense. (No one was going to pay to have a long-shot like me visit them.) I learned the difference between investment banks and commercial banks. I learned the types of problems of interest to these firms. Oddly, several people during these interviews commented on my Berkeley affiliation, and asked if I was familiar with BARRA, an investment consulting and technology firm located there.

The alumni interviews were also useful learning experiences. I had lunch with a Harvard PhD who worked as a biotech analyst for Montgomery Securities in San Francisco. She wanted to know why I wanted to be an analyst. I tried to hide my ignorance of what analysts did. My search was at a stage much too early to have such a specific goal. Navertheless, by the end of the lunch, I knew what analysts did and why I shouldn't become one. (It didn't play to my strengths in mathematical analysis, and included many requirements I lacked.)

Finally, I met an older Harvard alumnus working for Loomis Sayles in San Francisco. Our meeting was short. He was disappointed that I had only attended Harvard as a graduate student, not an undergraduate. He was also a traditional investor, with no contacts among quantitative investors. But he suggested that I talk to BARRA, as they were close by. And, he looked up the number in the phone book.

Having heard of BARRA from several people in New York, and now from this alumnus, I decided to contact the company. I found that Richard Grinold was BARRA's director of research and consulting, and the person to whom to send a letter of application.

BARRA's First Rocket Scientist

Unknown to me at that time, BARRA was one of the centers of the quant finance revolution. The name stood for Barr Rosenberg and Associates. Barr Rosenberg was a finance professor at Berkeley. More than anyone else, he had taken new academic theories, especially the Capital Asset Pricing Model (CAPM) and modern portfolio theory, and made them useful and accessible to practitioners. Consistent with that effort, BARRA expended considerable effort on educational activities, training clients on new ways of thinking about investing.

Just how revolutionary was all this? Rosenberg had appeared on the cover of *Institutional Investor* magazine in 1978. The cover illustration showed him sitting in a lotus position, in flowing robes, surrounded by much smaller men in suits bowing down to him. The headline read, "Who is Barr Rosenberg, and what the hell is he talking about?"

What he was talking about was modern portfolio theory. This approach consisted of several related ideas. Most importantly, what mattered to investors was their overall portfolio, not the individual stocks. So focus should be on the returns and risks of the portfolio. While the biggest challenge was forecasting returns, structurally this was straightforward. The return to the portfolio was simply the (weighted) average return to the assets in the portfolio. Forecasting portfolio risk was more complicated. It required knowing not only the risks of the underlying assets, but also their correlations with each other.

Harry Markowitz had received a Nobel Prize for developing this portfolio return and risk framework in the 1950s. Bill Sharpe shared that prize for combining this framework with equilibrium assumptions to develop the CAPM in 1964. Treynor, Lintner, and Mossin were on roughly the same track in the same era.

Rosenberg's critical contributions were to develop, for the first time, models for implementing these ideas in practice, and to place risk as the central framework for thinking about investing. BARRA developed risk models for financial markets. BARRA's primary business is selling subscriptions to these models. They forecast risks of individual assets, and, quite importantly, their correlations. The models did this by separating returns into common and idiosyncratic components. The common factors generated the correlations between assets. For example, the return to IBM consisted of a return to the computer industry, a return to large stocks relative to small stocks, as well as a component completely idiosyncratic to IBM, dependent on its CEO, its particular models and projects, and so on. The risk of IBM stock followed from the volatility and correlations of those components.

The power of this approach becomes most clearly apparent in analyzing IBM's correlation with other stocks. Its correlation with other computer companies follows from the fact that their returns all share that exposure to the computer industry. IBM's correlation with ExxonMobil follows from having a return factor in common (large stocks relative to small stocks), and the fact that the computer industry and the energy exploration industry returns are correlated.

These models provide a natural framework for thinking about return forecasting. Investors can forecast the factors, or a subset of them. Investors can forecast the idiosyncratic components of returns. Investors can carefully engineer their portfolios, pinpointing exactly where to take risks. Ex post, these models attribute realized returns to various components.

The commercial success of these models—and the ultimate buy-in from investors—followed from the intuitive nature of the factors. Investors already thought about industries and styles. Rosenberg quantified how these intuitive ideas connected to returns and risks.

More importantly, these models provided a framework in which to apply quantitative rigor to investing to engineer optimal portfolios. This framework facilitated the application of quantitative analysis to basically every investment problem. All of investing is optimally trading off return against risk and cost.

BARRA was a wonderful place to begin a career in finance. The firm's leaders, Andrew Rudd and Richard Grinold, were academics who fostered a learning environment. The firm was organized around a unifying framework for thinking about investing and applying that framework to all investments from stocks to bonds, domestic and international to activities from portfolio management to trading to hiring and firing portfolio managers. Research at BARRA differed from that at broker/dealers or investment managers in that it was less proprietary, since BARRA sold risk models and not actively managed investment products. The firm ran well-known annual research seminars at

Pebble Beach, California—part client education, part presentation of new results from quantitative analyses of investment issues.

I joined as the firm's first "rocket scientist," someone with extensive training in science and none in finance.[2] When Richard Grinold hired me, he also invited me to sit in on the graduate seminar that he was teaching at Berkeley.[3] My first months at the firm included in-depth work on a particular project, as well as a survey of seminal papers in academic finance.

My project was to research an improved model for interest rate options. For example, the U.S. Treasury had long issued bonds with embedded options, allowing them to pay back investors once the bonds were within five years of maturing. Corporations issued similar bonds. If the bonds paid 12 percent interest, and rates had fallen to 8 percent, the Treasury could refinance at a lower rate. A bond that pays 12 percent interest for six years looks very different from a bond that pays the same interest for only one year, at which time it matures. The six-year bond is worth more, is more sensitive to interest rates, and is more volatile. BARRA's bond model needed to correctly analyze such bonds—in particular, how such a bond transitioned from six-year maturity to one-year maturity as interest rates fell. In general, the impact of these embedded options increased as interest rate volatility increased.

Years earlier, Barr Rosenberg had developed a model to analyze interest rate options. It was ahead of its time when developed, but was designed assuming that these embedded options had only a marginal impact on value and risk.[4] In the Volcker era, these embedded options suddenly had very significant impacts on bonds, and required better models. Richard Grinold had done some preliminary work on replacing the existing BARRA model with a model developed by Cox, Ingersoll, and Ross.[5] My project was to continue this work and to implement the new model if it proved up to the task.

As I began my new career in finance, I remembered a comment from Bill Press, my advisor in graduate school. He told me that undergraduates learned from their professors. But the role of a graduate student was to teach their advisor new things. I set myself the goal that, by the end of my first month, I should tell Richard something he didn't already know. I was clearly unable to tell him something new in finance after one month in the field, so I tried to show him some ways to improve the computation speed of the calculations. I also set myself similar goals for every month, ultimately aiming to tell him new things in finance.

I also remembered what Winston Churchill told his cabinet members upon becoming prime minister in 1940: Important communication should always be in writing and limited to one page. That appeared sufficient to win the Second World War, and so was a good benchmark. In my case, I wrote up reports every month (succinct, though usually four to five pages), describing my progress and results, as well as my planned next steps and current questions. The act of writing clarified what I really understood. And the reports were a terrific way to work with Richard. He could see progress, and very efficiently provide me with guidance. That was the beginning of a working relationship now in its twentieth year.

The interest rate option project ultimately fed into BARRA's bond-risk model. Instead of the industries and styles used in the equity models, the bond model factors included interest rates and sectors. The interest rate option model helped determine exposures to interest rates for all assets with embedded options. For example, ignoring the embedded option, a callable bond might be partly exposed to movements of short- and intermediate-maturity interest rates, and partly exposed to movements of long maturity interest rates. But if rates had dropped significantly, and the issuer was almost certainly going to exercise an option to refinance the debt, the bond maturity would effectively be much shorter. The bond might not have an exposure to movements of long-maturity interest rates.

My work on that first project provided me with considerable insight into how interest rates moved, and how no-arbitrage arguments led to option pricing models. It also helped me begin to understand how interest rate option models fit into the larger-risk model, and how investors used that model. But my knowledge of finance at that time was an inch wide and a mile deep. I knew the Cox, Ingersoll, Ross model in considerable detail. And that was almost all I knew in finance. In October 1987, after eight months at BARRA, I attended my first Fixed Income Research Seminar at Pebble Beach. I presented an optional session on my work. Richard instructed everyone to only let me talk to the most technically sophisticated clients. Although I could discuss technical details about option models, I would have struggled with more general questions.

Success in finance would require me to acquire much more intuition and insight into the basic problems faced by investors and to build up a toolkit for analyzing those problems. My approach was very different from what a formal education in this area would involve. Basically, I tried to master specific areas—like interest rate options—and then branch out and master new areas, through new projects. It was a patchwork quilt approach. I hoped I could build enough sufficiently large patches that they would eventually form a substantial quilt.

I first branched out into building other components of a new and improved bond risk model. I worked on understanding the common factors impacting bond returns. Beyond interest rate effects, did bonds of similar quality ratings all move together? What about bonds in similar sectors? How do we apply interest rate option models to mortgages, where the refinancing option is similar, but the mortgage represents a pool of investors, each with their own differing personal circumstances? And given a history of rate and sector and quality movements, how do we best forecast their future volatilities and correlations?

Those projects all mainly supported the development of an improved BARRA model. But I also worked on two special research projects aimed only at improving investor insight into particular issues of interest. The first project investigated the value of *convexity*. There was a quant argument (first discussed, I believe, by quant Stanley Diller, a Columbia economist who moved to Wall Street in the 1970s) based on a simple Taylor series of how bond prices, P, depended on yield, y, especially as the yield moves away from an initial value y_0:

$$P(y) = P(y_0) + \left(\frac{\partial P}{\partial y}\right) \cdot (y - y_0) + \frac{1}{2}\left(\frac{\partial^2 P}{\partial y^2}\right) \cdot (y - y_0)^2 + \cdots$$

$$\begin{aligned} \textit{return} &= \left(\frac{1}{P}\right) \cdot \left(\frac{\Delta P}{\Delta y}\right) \approx \left(\frac{1}{P}\right) \cdot \left(\frac{\partial P}{\partial y}\right) \cdot (y - y_0) \\ &\quad + \frac{1}{2}\left(\frac{1}{P}\right) \cdot \left(\frac{\partial^2 P}{\partial y^2}\right) \cdot (y - y_0)^2 \end{aligned}$$

The second derivative term, $(\frac{1}{P}) \cdot (\frac{\partial^2 P}{\partial y^2})$, is the convexity. The first derivative term, linear in the change in yield, could give rise to positive or negative returns, depending on whether yields moved down or up, respectively. But the convexity term, because it multiplies the squared change in yield, contributes positively to return whether yields move up or down. So convexity looks like an unambiguously good thing.

There are two problems with this. First, if everyone in the market understood this phenomenon, and was willing to pay a higher price for bonds with more convexity, that higher price should eliminate any free lunch here. Second, this analysis implicitly assumes that yield is a simple variable. In fact, yield is an average interest rate over the bond's maturity. The yield could change because short rates moved or because long rates moved. But those two movements would lead to different bond returns.

My colleague Roland Lochoff and I tested investment strategies that bet on convexity to outperform the bond market. Each month, the strategy would overweight high-convexity bonds and underweight low-convexity bonds while hedging out exposures to other factors, and minimizing risk. It was a classic quantitative analysis of a possible investment strategy. We built portfolios that optimally traded off expected return against risk, assuming returns proportional to convexity, and used the BARRA bond model to forecast risk. Our resulting paper[6] showed that the market appears to correctly price convexity. Convexity-based strategies do not beat the market.[7]

Beyond this specific result, the project taught me a new methodology for analyzing this type of investment question. And, it began my focus on researching investment ideas that attempt to beat the market. This initial patch would eventually grow to be a significant part of the quilt.

The second special project looked at a risk worrisome to bond investors in the wake of the leveraged-buyout (LBO) of RJR Nabisco. There, company management borrowed huge amounts to buy out existing shareholders. That increased debt burden significantly lowered the credit quality of the company, decreasing the value of existing bonds. The LBO caused a significant transfer of wealth from bondholders to stockholders.

The research opportunity here was to learn two new (to me) financial theories and to combine them to analyze the problem. In 1958, Franco Modigliani and Merton Miller, both Nobel laureates, had shown that under idealized circumstances the value of the firm was independent of leverage. Shifting between equity and debt just shuffled claims on firm earnings between equity holders and bond holders. They later showed that in the presence of taxes, a shift from equity to debt could increase firm value by lowering the effective tax rate. Robert Merton, another Nobel laureate, developed a model applying option theory to value bonds as a function of the debt/equity ratio. By combining these models, I analyzed the incentives and the impact of LBO's on bond holders.[8]

This project taught me new ideas, indulged my personal passion for combining different ideas to solve new problems, and exposed me to yet another investing topic of interest. It represented another project applying finance theory to understand a practical problem.

Those early years at BARRA were extremely exciting. The work was fascinating. I was experiencing the steep part of the learning curve, perceptibly expanding my knowledge on a daily basis. It was possible to make significant progress on problems with a few months of effort, and investors genuinely wanted to know the answers. My colleagues and supervisors were supportive, and extremely smart. I began to see that those physicist conceits were wrong. I was meeting people as smart as anyone I knew in physics, and some of them had never even considered physics. I worked on nonphysics problems that fascinated me.

Other preconceived notions proved false as well —for example, the idea that in physics you belong to a collegial community of scholars, with society quickly recognizing your contributions, while in finance you swim with the sharks, enduring backstabbing and greed. Or the idea that physics benefits society, while finance exploits widows and orphans. To the contrary, I found people in finance to be more supportive, and quicker to recognize contributions, than those in academia. I could also see much of the work in quantitative finance aimed at improving the lives of investors—that is, making a positive contribution to society, and one more concrete than anything in the recent history of physics. But these preconceived notions remain strong amongst some physicists. Years later, attending my 25th college reunion, I was shocked when I ran into a former physics undergraduate, now a physics professor, who told me he had heard that "I'd gone over to the dark side."

There was no question, though, that finance differed from physics. It is not an exact science, and involves few immutable laws of nature. As former physicist Al Slawsky put it in the early 1990s, "It was a bad year for value investing. In my former life, there was never a 'bad year' for gravity." The physicists who succeeded in switching to finance all quickly understood they were doing something different.

That said, a physics background stood me quite well in finance. Wall Street had good reason to hire physicists. Graduate school in physics provides a rigorous training in quantitative problem solving. Even if finance involved some different mathematical techniques, that training made it easy to pick them up. Also physicists learn to solve problems by identifying the 90 percent of the problem to ignore, and focusing on the 10 percent that matters. We do not typically try to solve problems exactly, but aim for capturing the most important elements and getting close. That skill is, if anything, even more important in finance, where there typically are no underlying laws of nature, and the problems are mainly about reasonably approximating reality. Finally, physics, and especially theoretical physics, is all about abstractions. Operating at a high level of abstraction makes it easy to switch to a different manifestation of the mathematics, and to think about finance problems in the broadest possible terms. After thinking about infinite universes, it's easy to deal with the most abstract elements of finance.

Around this time, my responsibilities at BARRA expanded beyond bonds. I began working on a project to apply these ideas of risk, return, and cost to equity trading. The portfolio manager trades off these three components, deciding which stocks to buy and sell to optimize the portfolio. The trader has a different problem. The portfolio manager provides the trader with the list of stocks to buy and sell. The trader must decide how to optimally schedule those trades (i.e., how much of each stock to trade each day). The optimal schedule provides the best tradeoff between returns, risk, and cost, with an assumption that costs increase as trading speeds up. With equity trading specified as an optimization problem, we could attack it with BARRA technology. This view of trading appeared radical to most traders at the time, since they focused on trading single stocks, not portfolios of stocks. One key research challenge was to develop models of exactly how costs changed as trading sped up.

The equity trading project was not another research paper, but the development of a new product for BARRA. As such, it evolved into a multiyear, multiman-year effort to develop the required component models and combine them. It provided me yet another opportunity to expand my sphere of knowledge well beyond that original interest rate option model. It also provided further evidence that the risk, return, and cost framework applied very generally to problems in finance.

Active Portfolio Management

In 1990, Richard Grinold offered an internal course at BARRA, combining academic theories and seminar presentations to sketch out a scientific approach to active management. Richard's goal was to turn this into a book, and he offered the course as a way to make progress in that direction. But he eventually realized that he wouldn't be able to do this on his own, and he asked me to join him in the effort.

Our goal was to put together a rigorous framework for scientific active management—the science of investing—presenting the view of investing as optimizing return against risk and cost, and developing component models, tools, and analyses. We wanted to provide the analytical underpinnings for a new class of active managers, who apply rigorous analysis and process to try to beat the market.

Although many of the components of this framework were known, even if not meticulously derived, there existed no work that pulled the components together. The major reason for this was that finance academics—the most likely potential authors of such a work—had spent two decades arguing about whether markets are efficient. They led the move away from active management, toward indexing. Only very rarely had any of them asked what to do if you have a valuable insight.[9] The contribution of our book, *Active Portfolio Management,*[10] was to analyze that question in detail.

To start, we moved away from the abstract concept of *the market*, so often studied in business school, to the more concrete performance benchmark. Active management is about beating a benchmark, with return relative to the benchmark defined as *alpha*. Any analysis of active management must hence start with that benchmark. Defining active management as an optimization problem—trading expected return against risk—naturally connects returns to portfolios. So any set of expected returns leads to a unique optimal portfolio, and a portfolio assumed optimal leads to a unique set of expected returns. The benchmark portfolio itself naturally leads to a set of *consensus* expected returns (because the benchmark is the consensus portfolio). These consensus returns look like a Capital Asset Pricing Model result: They are proportional to covariances with the benchmark. Holding a portfolio that differs from the benchmark—almost the definition of active management—implies a set of expected returns that differ from the consensus. The job of active management is to forecast active returns—returns relative to those consensus returns—and optimally trade off active return against active risk, the volatility of those active returns.

Given that framework, investors will maximize active return relative to risk by maximizing their information ratio (*IR*), their ratio of active return (*alpha*) to active risk. The information ratio is effectively a measure of consistency. Investors seek consistent positive active returns. High *IR* managers deliver positive active returns more consistently than low *IR* managers.

With the importance of the information ratio in mind, Richard Grinold's "Fundamental Law of Active Management"[11] shows that information ratios depend on the product of skill and breadth. Skill measures the investor's edge in every investment decision. For example, how often do their stock picks outperform the benchmark? Breadth measures the number of available independent decisions. In this same example, how many stocks do they follow, and how often do they analyze new information? High information ratios require an edge, and the opportunity to diversify by applying that edge many times. Due to the fundamental law, quantitative active strategies tend to take many small bets as opposed to a few concentrated bets. The goal, based on this framework, is to deliver consistent performance.

Beyond these basics, the book provided considerable guidance into how to build and test investment strategies, how to properly optimize portfolios, how to model and account for transactions costs, and how to analyze performance ex post. The book did not provide alpha ideas—as such, ideas only work if the market doesn't already understand them.

Active Portfolio Management has played an important role in legitimizing the science of investing. While the consistent investment performance of quantitative managers like Barclays Global Investors was critically important, the flow of institutional assets into quantitatively managed investments also required intellectual legitimacy, which *Active Portfolio Management* has helped provide.

Barclays Global Investors

In 1991, Richard Grinold became executive vice president of BARRA and, fairly soon after that, president. When he became executive vice president, I became director of research, a role I held for over seven years. Richard left BARRA at the end of 1992 to join Wells Fargo Nikko Investment Advisors, which later became Barclays Global Investors (BGI).

My years as director of research at BARRA were fantastic in terms of my intellectual development. I worked on many interesting projects, including supervising the development of new equity, fixed income, and trading models. Beyond researching new BARRA products, I participated in numerous research studies that appeared as seminar presentations and subsequent articles. Andrew Rudd and I analyzed that ultimate question for investors: Does historical performance predict future performance?[12] Our contributions included better quantitative apples-to-apples comparisons of managers, and an analysis of whether any perceived persistence of performance could lead to an outperforming investment strategy. Our answer was no. Analyzing that data led to a subsequent paper, "Bond Managers Need to Take More Risk,"[13] showing a problematic mismatch between manager consistency, risk level, and fees. Typical bond manager fees were high relative to their potential performance, and many products weren't configured to deliver positive performance on average. I also wrote about why it was so easy to datamine—to find interesting patterns in historical data—but patterns that would not persist in the future.[14] This was closely related to the statistics of coincidence—why coincidences seem to pop up so often.

The people with whom I interacted were also very interesting. I learned many things from my colleagues and superiors, and the members of my research group. I greatly enjoyed talking with clients and prospects at our research seminars, and in other visits, including a memorable visit with Fischer Black at Goldman Sachs Asset Management.[15]

I also remember four separate meetings one day at Morgan Stanley, during another trip to New York. I met about a dozen people in all, and only one was not a former physicist. One in particular described moving from the canceled Superconducting Supercollider (SSC) project to Morgan Stanley, and rejoiced in working somewhere that his superiors cared about what he was doing.

In spite of these many fascinating projects and interactions, by 1998, after spending years telling others how to manage money, I wanted the experience myself. I would be giving up a public platform I enjoyed to concentrate on proprietary research, but I wanted new challenges. It was natural to join Richard Grinold at BGI, the firm that had been among the first to commercialize the concepts in *Active Portfolio Management*.

Since joining BGI, Richard had substantially enhanced its active investment research, and the firm wanted to dramatically increase its active management business. I joined as director of research for Japanese equities. My focus on Japan was deliberate. I wanted a narrow focus for my entry into a new (though related) field. I had a long-standing interest in Japan and the Japanese markets. And BGI's Japan strategies weren't working well.

Over the next two years, I helped build up the Japan investment team, and we developed a new generation of proprietary equity return forecasts. We established close working relationships between researchers, portfolio managers, and traders—especially impressive given that team members sat 5,000 miles and eight time zones apart, with different native languages—the model for all the equity teams at BGI. I found it very rewarding to build teams and see them succeed in a highly competitive activity.

As the Japanese products started to generate great investment returns, I switched to heading BGI's active equity business in the United States. In early 2001, our biggest challenge was not investment performance, as our equity products had impressive track records. Instead, it was convincing prospects that we were more than just a big index fund provider. In part, this was just BGI's problem. But in part it also spoke to the very limited legitimacy of quantitative active management. Over a two-year period, my colleague Scott Clifford and I devoted considerable amounts of our time toward changing these perceptions. As investors began to notice how BGI and other quantitative managers had managed to deliver on return promises while controlling risk over long periods of time, perceptions changed. Significant new investments flowed in.

At that point, I moved on to become global head of equity research. Consistent outperformance requires constant innovation, and that requires as large and creative an investment team as possible. I have spent much of my time recruiting smart and creative people to join us. And over that time, with the help of others, we have assembled an amazing collection of many of the world's best quantitative financial minds. The BGI research team would count as one of the top academic finance departments, yet instead of publishing research, the group is 100 percent devoted to generating alpha. This is great for our clients, and for us. It helps make BGI a wonderful place to work.

Although much of my work at BGI has been proprietary research, or management, I have researched some investment issues we were willing to share broadly. These efforts included analyzing the efficiency of long-short investing[16] and analyzing capacity of investment products.[17]

My experience at BGI has been a culmination of my path from physics into quantitative finance. I learned finance in patchwork fashion, and then participated in developing and systematizing the quantitative approach to active management. Moving from BARRA to BGI offered me the opportunity to put these ideas into practice, where our investment and business success provided concrete validation for the science of investing.

The Future

At this time, the art of investing has substantially evolved into the science of investing. Quantitative active management is no longer an obscure subfield of investing, but a fast-growing and significant enterprise. There is substantial momentum behind it.

That would argue that the revolution is over. And certainly much of the relevant theoretical underpinnings are in place. At the same time, there exist even today significant issues in need of rigorous quantitative analysis. Present-day debates over liability-driven investing prove that fundamental issues remain unresolved among investment professionals.

Other areas worthy of further research include hedge funds and, more generally, optimal leverage and shorting.

Between these remaining fundamental issues and the never-ending competition amongst active managers for outperformance, the need for quants has never been greater. There is plenty more to do in this challenging, interesting, and rewarding field.

Chapter 3

Gregg E. Berman

Strategic Business Development, RiskMetrics Group

I would not call myself a *quant*. In fact, I would hesitate to call anyone a *quant*, for a couple of reasons. The first is that on Wall Street the term can sometimes be used pejoratively, as in "He's *only* a quant," rather than a true businessperson. The second reason is that I don't think there is such a profession as a *quant*. To be sure, one can certainly be quantitative. But that's an attribute of a person, not a career description.

For instance, as a quantitative person, one might be employed as a pharmaceutical engineer, a classical-era musicologist, or even a developer of financial analytics. But quantitative people are also Olympic bobsledders, late-night talk show hosts, and politicians. The only counter-example I can think of is the group of women who, as part of the 1940s Manhattan Project, helped solve essential numerical problems involving the development of the first fission bomb. I suppose that, because their sole task was to perform repetitive calculations in a somewhat isolated fashion, one can call them *quants.*

So please allow me to take some liberties with the title of this book and split it into two parts: "How I became a quantitative person" and "How, as a quantitative person, did I eventually find my way to Wall Street?"

A Quantitative Beginning

The first question is easy to answer. I was just born that way. According to my aunts and uncles, I was able to count before most babies read. Apparently, much to the amazement of onlookers, I had a habit of reciting elevator floor numbers as they winked on and off in the various Queens, New York, apartment houses where most of my relatives lived. I obviously don't remember anything from when I was six months old, but if I was indeed reading aloud floor numbers as we rode up and down, then I'd like to believe I must have been computing our rate of ascent.

The second question is, of course, the topic at hand, and through its answer I hope to provide some insights into the lessons I've learned, what I think I did right, and what I think I did wrong. And, if I'm lucky, you will see some patterns or similarities with the choices you have faced, or may face in the future.

I never set out to have a career in finance. I had always enjoyed math and science, and whenever we took those somewhat silly vocation tests in school, I would try to rig my answers so they would indicate my proclivity to be a scientist. Curiously, they always predicted my ideal profession as law or acting. Nonetheless, in my first year of junior high school I announced to the world (or at least to my teachers) my intention of becoming a physicist and attending MIT. In 1980, after programming on school mainframes for a number of years, I received my first microcomputer, a Radio Shack TRS-80 Model 1. All of the programs I wrote involved calculating which types of quarks were found in which types of subatomic particles. There is no better definition of the word *nerd*.

But alas, there also seemed to be some foreshadowing in my hobbies. I wrote a checkbook balancing program, sort of like Quicken, but a thousand times slower. Remember that in 1980 floppy disks were just coming out, and personal hard drives were six or so years away. Nevertheless I found enjoyment in working with financial calculations. Perhaps even more relevant was my development of a betting game that simulated horse racing. It probabilistically determined how far each horse would advance per gallop according to randomly generated payout odds. If instead of simulating horse strides I had been simulating stock prices, I would have been called a financial engineer rather than a geek.

The second biggest influence on my ultimate choice of careers came from my parents' financial status during my most formative years. No, they were not Wall Street aficionados. Just the opposite—they sold toys and stuffed animals in various Long Island flea markets. And so, throughout my junior and senior high school years, I worked the market every Saturday and Sunday, as well as weekdays whenever the flea market was open in the evenings. During busy seasons (Valentine's Day, Christmas, etc.) I teamed up with my father to sell our wares, create displays, or trek into Brooklyn in an old, seatless Ford van to buy odd lots of toys at questionable wholesale outlets. When the market was slow, I worked for other vendors, selling anything from pretzels (75 cents each) to frozen ices (also 75 cents) and plastic clothing hangers (eight for one dollar).

Though I know the economic pressures my parents faced during those years were tremendous, I myself had a great time wheeling and dealing. I learned about what it really meant to be on the hook for closing a sale. I learned about people, about negotiation, about presentation, and about the disparities of what is delivered versus what is promised. In summary, I was getting a true Wall Street education without even knowing it. Moreover, the unlikely pairing in my mind of academic physics with real-world business issues made a huge impression. Looking back on this period of my life, I can easily see how the congruence of these two forces directly shaped my future and strongly influenced my career path.

Patience is not a virtue to which I subscribe and, at the beginning of my junior year of high school, I informed my guidance counselor that I intended to double up on classes and graduate a year early. Surprisingly, almost everyone tried to convince me of the importance of staying around for my senior year in high school. After all, how could I miss out on the senior prom, and the potential camaraderie of varsity sports?

Since I thought high school was the moral equivalent of purgatory, and math club did not count as a varsity sport, I did not heed their warnings. The only ones wholly supportive of my decision were my parents and a key teacher who ran our school's computer club and with whom I had become good friends. Together, we founded a top-notch computer team that became somewhat famous throughout the New York area for winning a plethora of programming competitions. It's amazing how influential the right teacher can be.

By the end of my junior year I had almost all the credits I needed to graduate early. In spite of completing advanced calculus, physics, and chemistry, I apparently did not have enough gym credits. I guess it was imperative that I prove doubly proficient in dodge ball lest my professional goals should ever require me to evade bright red rubber spheres hurled at me with great speed. Nevertheless, I completed everything needed to graduate by June, got a summer job at a computer store selling Commodore 64s, and started in the fall of 1983 as a freshman at MIT.

I absolutely loved "the Tech." I every physics class they offered, and even a few they didn't. On weekends and over the summers I worked on campus at the Center for Space Research helping to design and build X-ray detectors for NASA's Advanced X-Ray Astrophysics Facility (never launched). My research was split between hardware and systems, but my thesis focused on modeling photoelectron flow inside our prototype charge-coupled devices. Today, high-resolution CCDs are found in simple $50 digital cameras; but twenty years ago this was real cutting-edge stuff. Once again I found myself coding advanced simulations, though this time in C on UNIX platforms instead of in Basic on a TRS-80.

Every day at MIT further reinforced my self-decreed providence, and all signs pointed to physics, at least those signs I was willing to acknowledge, even as other interests hinted at different possibilities. A great example occurred during my third year at the Institute, in which I became president and national chairman of a student space-advocacy group called SEDS. What's a space group good for if it doesn't have a telescope? So I petitioned the student counsel for funds, and received a laugh when they found out that the cost of the 8-inch Celestron we wanted was about $3,000. Instead, they told us to try to raise money doing what all the other student groups did—holding bake sales. But since most of these events netted $50 or less, and none of us could bake, I needed to find another way.

About six months later on an "adventure tour" to a suburban mall outside of Boston (this is my definition of being adventurous at MIT), I ran into one of those ubiquitous pseudo-store kiosks. But this one housed a guy selling space-art, such as $10 posters of nebulas, the rings of Saturn, and of course everything ever printed regarding Star Trek. Guided by my market background, I introduced myself and convinced him to lend his entire stock to our student group on consignment for sale at MIT. He agreed, so we held a three-day sale in the main corridor of MIT under the great dome, and by the time it was over every student on campus had at least one of our posters hanging in their dorm room, and we had $3,000 in small bills. Posters of the Starship Enterprise and pictures of fantasy unicorns—who would have guessed they would be so popular at MIT? Our take of the proceeds was $1,500. Check in hand, I went back to the student counsel with my fund-raising results. The president put me on the counsel, matched our funds with a second $1,500 check, and SEDS bought our first telescope by the end of the week. And what did this have to do with physics? Absolutely nothing. Wonder what that meant?

Putting It to the Test

I graduated from MIT on schedule in the spring of 1987 and, having no thoughts other than to pursue my PhD, I found myself at Princeton that fall. I was surrounded by more Nobel-prize-winning physicists than I had ever imagined. This, after all, was the land of Albert Einstein and I myself was following in the footsteps on the great physicist Richard Feynman (he had also gone to MIT as an undergraduate). By October, I was as well-entrenched in class work as in research at Princeton's premier radio-astronomy lab, where I was studying the nuances of millisecond pulsars. I had arrived.

In fact, the intellectual aura at Princeton was somewhat intimidating; and as a first-year graduate student I was keenly aware some professors were known to grill students not just in class, but also during afternoon tea breaks, and even in the courtyard. My first such experience came that fall when simply walking down a corridor outside the astrophysics lab: I passed a person without much notice when suddenly I heard him say, in a very thick, slow, European accent, "Excuse me—may I ask you . . . a question?" I cautiously turned to find myself face-to-face with Professor Wigner, one of the university's most prestigious professors and considered among the founding fathers of nuclear physics. There were even equations and theories that bore his name.

First came the cold sweat, then the shakes, and finally an inability to speak. What could Professor Wigner possibly want to ask me? Nuclear physics was something I had not yet studied in great detail. I was sure I was going to fail this ad-hoc test, proving an embarrassment to myself, to my colleagues, and to everyone in New York/New Jersey metropolitan area. Yet I managed to respond with a squeaky "Y-Y-Yes P-Professor Wigner, what can I help you with?" After what seemed an immeasurably long pause, he asked, "Do you know . . . when I call a long-distance number from my lab phone . . . do I dial a nine first?" A broad smile broke out on my face as I said with confidence, "Yes, Professor Wigner, you do dial a nine first, followed by a one, the area code, and then the telephone number." He smiled back and said, "Thank you, young man, you have been most helpful." I had passed my first test, and though it was not really related to physics, it did illustrate just how important having the right information at the right time can be.

That January, my research professor and I went to world-renowned Arecibo Observatory in Puerto Rico to study pulsars at their 1,000-foot radio telescope by using the custom electronic filters and specialized software algorithms I had been building. This was to be the culmination of everything I had worked for—a holy grail of physics.

I spent a total of a month on that mountain, first with just my professor, and then with a pair of Russian scientists who were not at all impressed with our American-style electronics, computers, or telescopes, but were fascinated by the ease at which I punched holes in note paper to fit into a three-ring binder. Apparently they had no problem making do with ex-Soviet era technology, but they never moved beyond the single-hole punch. As a fair swap, we gave them a three-hole punch to smuggle back into Russia and they gave us a few bottles of homemade vodka.

So there I was, hanging out on a mountaintop in Puerto Rico—playing with a 1,000-foot telescope, watching the stars, doing shots of Russian vodka. And I thought to myself: "I really hate this—I'd rather go home."

Huh? Where in the world did that come from? How could I possibly not like what I was doing? Wasn't this what physics was about? Wasn't this what I always wanted? Wasn't this what I was destined to do? By the time I got back to Princeton, I was thoroughly depressed and confused. To make matters worse, I had to prepare for the physics department's preliminary exams, which all first-year graduate students take at the end of May. This two-day written test is supposed to check your preparedness for beginning graduate work. But I spent so much time grappling with my internal conflicts about physics as a career that I did not spend much time studying. That year's physics prelims were unfortunately quite difficult, and seven out of 30-some-odd graduate students failed the test. I was one of those seven.

Failing prelims was not the end of the world, though it seemed that way to me. You could take them again in the fall, but that was your only chance. Fail that and you're asked to leave. Which left me with a real dilemma—do I study for the exams in order to continue with my degree, which I'm not sure I really want, or do I pack it up and move on? In the end, I decided that the only thing I hated more than being depressed about physics was failing. If I were going to leave physics it would be on my own terms and after a successful run, not because of some silly test. So I studied my brains out in what has become known at Princeton as the Summer of Despair. We lived in Princeton's less-than-quaint Graduate College dormitory with no air conditioning and barely edible cafeteria food. That summer the temperature in New Jersey broke 100 degrees for something like 21 days in a row.

Summer ended. Fall arrived. I retook the prelims and passed with flying colors. But rather than celebrate, I had to start studying anew for the even more intense set of exams that all physics students take at the end of their second year. These are Princeton's official qualifying exams for PhD candidacy, known as *generals,* and they consist of three days of written tests, two days of orals, and a lab. I therefore split my time between studying for generals, continuing my research in the radio-astronomy lab, and teaching an undergraduate lab class (affectionately known as Physics for Poets). Throughout the time I became increasingly disillusioned and unhappy. I studied on my own and with groups of other students; but I did not bother attending classes, which I found to be a waste of time. And I let everyone around me know it, including my professors. What made matters worse was that I couldn't figure out why I was still unhappy, and the faculty was wholly unsympathetic.

In fact, my relationship with some professors became difficult when they realized I was not abiding by many of the unwritten rules developed at Princeton. For instance, in an attempt to find some extracurricular fulfillment, I started an intradepartmental newsletter in which graduate students from each lab outlined their current research. The idea was to promote cross-lab communication and reduce redundancies if two labs were trying to solve the same problem—something I had become keenly aware of during my research. I was particularly pleased with our first issue of the Jadwin Chronicles, and feedback from the other students was extremely positive. To my surprise, I was called into the department chairman's office after he learned of the publication, but rather than offer enthusiasm, he read me the riot act. It was apparently unacceptable for a graduate student to put out a newsletter. It was not the content of the newsletter itself that was the problem. It was the fact that I felt I had the time to spend on it. It obviously meant I wasn't working as hard as I could in the lab or on my studies, so I should stop doing anything not related to my research.

The end of the year came. I sat for generals. I passed the test. I have no idea how, except to say that my displeasure with the whole program gave me a perverse desire to succeed at any cost. However, the experience took quite a toll on my well being and I felt I needed a summer off before diving back into lab work. I also wanted to see what the world outside of academics was like, so I took my first step by applying for a summer internship at General Electric Astrospace, located only a few miles from the university. I got the job by simply calling its HR department, introducing myself, and asking if they had anything available for the summer. They hired me on the spot without knowing anything but my first name and that I was a graduate student studying physics at Princeton. The position was to be in Astro's Conceptual Design Department, responsible for simulating various aspects of the Mars Observer spacecraft to be built and launched a few years hence.

A Martian Summer

I was in absolute heaven. My dream job at last, especially considering it came with an annualized salary equal to $33,000, which seemed enormous compared to my base $12,000-per-year stipend from Princeton. By the end of my first week I was deep into a few hundred thousand lines of ancient Fortran code trying to decipher why the effects of the Earth's precession were not properly taken into account when tracking the coordinates of Phoebos and Deimos (the moons of Mars). This was important, since we were trying to simulate how often the moons would block the main antenna of the spacecraft if the antenna boom was shortened by three inches to save weight.

So what did I learn from my summer experience? I learned I liked the outside world. I learned I liked working for a commercial company. And much later, I learned that once the billion-dollar spacecraft I had worked on finally reached Mars, it exploded. Oh well. But most importantly, I learned never to tell your professors and friends that you are taking off the summer to work at GE Astrospace. I was an instant pariah. The hand-written note I received from Professor A. was one of the most vicious letters I had every read. How dare I take leave from the temple of physics! Who the hell was I that I should take off a summer and abandon professors who desperately needed lab grunts, oops, research assistants? Most of the letter I can't recite here, due to the colorful nature of the language.

Physics on Trial

By the time I returned in the fall, I was being called to task by a new chairman, who demanded that I pick my thesis topic and get started on my doctoral research. I had already decided I did not want to pursue radio astronomy. What then? Well, my experience over the summer helped me come to grips with my problems with physics and general unhappiness. It wasn't physics that I disliked. In fact, I loved it, and still do. Rather, it was the academic environment itself that I really disliked. It was how physics was done, how professors and research worked (and did not work) together. I can easily write a treatise on the problems with modern academic research. But let's just say I decided that the day-to-day workings required of the academic life were not to my liking, even if the ultimate pursuit of physics was.

At the risk of sounding preachy, this was a life lesson I feel fortunate to have learned so early in my career. Simply put, I realized it was much more important to be true to yourself than to remain loyal to a preconceived notion. In my case, the notion was that I had to be a physicist. The mere fact that I had worked so long in attaining that goal seemed to validate the concept. But of course that's a completely circular argument, and just because you spend 10 years on something doesn't mean you have to spend another 10. What it does mean is that you must make sure you are enjoying the first 10 years while they are unfolding, because the next 10 years never really arrive. Too many people I meet are trapped in jobs they no longer like, or never liked in the first place, because they feel loyal to some idealized concept of the career itself and are unwilling to admit the realities of the job may not, and may never be, fulfilling.

With that in the back of my mind, I decided to pick a thesis topic to maximize my fulfillment on a daily and weekly basis, regardless of the field itself. After all, can one really be loyal to experimental particle physics versus atomic physics? All the problems are interesting; so as long as you like what you doing, you're fine. Well, apparently not, according to my professors. They insisted I pick a topic based on my passion and adherence to a particular field at any cost. I guess you are not allowed to generically like baseball or football. You must pick a team, become a fanatic, and trash all the other teams. If you don't, then you can't really be a sports fan.

The next three years came and went. I joined a biophysics lab at the insistence of the chairman, but left after six weeks because I never really was interested in biophysics. Instead, I transferred to the nuclear physics lab in the basement. They had a 30-year-old cyclotron and even older spectrometers. I could spend countless hours in the machine shop rebuilding experimental equipment and even more hours writing simulation code of types never before seen in the department. That was my definition of a good day. Still, the bad days outnumbered the good ones. I was forced to switch experiments twice in midstream. The first time was because my advisor (a sixth-year assistant professor) did not get tenure and left Princeton. The second time was because the new professor I worked for was himself quite new and just finding his way. By then, I realized the only way to ever get anything done is to work for the boss. I therefore picked an experiment that was run by the head of the lab himself—one of my better decisions at Princeton. He made sure I got the equipment I needed and beam-time required. Another good lesson learned: Work directly for the boss whenever possible!

A Twist of Fate

Then, while knee-deep in my experiment, something truly fortuitous occurred. A Wall Street recruiting agency (a head-hunter) placed a random call to the secretary of our department asking if anyone was interested in working for a boutique hedge fund. She had no idea what that meant and politely told the recruiter to try another department. But wait: my nonconformity regarding the physics department was quite well-known to most of the staff. I even found out I was sometimes mentioned at faculty meetings, having been deemed a *rabble rouser*. No kidding, that was the term they used. Now being known as a rabble rouser might be undesirable, but spun correctly, it implied I was on the lookout for interesting ideas and novel opportunities. In general, you make your own luck by putting yourself in situations to take advantage of unknown possibilities and undefined potential. In my case, that luck was the departmental secretary knowing my situation very well; she therefore gave the recruiter my phone number and told him to give me a call.

That afternoon my phone rang, pleasantries were exchanged, and the recruiter told me about a boutique hedge fund looking for doctoral candidates in physics to join their research team. The year was 1992. I had no idea what he was talking about, and I still had a year of thesis work left. I took his name, asked him to send me a business card, and promised I would call him in a year when I was closer to completion of my degree. He hung up disappointed.

Another year passed. I had nearly finished my experiment and had to make a decision about next steps. My professor was adamant about me staying on as a postdoc and working deep in the bowels of an Italian mine performing delicate neutrino-physics measurements somewhat related to my thesis. This did not sound even remotely appealing. I could, of course, consider other postdoc positions, but the real question was whether to stay in physics itself. After coming all this way, how could I possibly give up now? I don't think I had a full night's sleep all that year, and not just because some of my experiments were run at 2:00 A.M. Moreover, I knew that taking a postdoc position was truly a commitment toward becoming a professor. If I was going to leave the field, I had to do it now.

A Point of Inflection

In the end, the decision was relatively obvious because I didn't have to decide the "big question" of whether I wanted to be a physicist. Instead, I started by deciding whether I thought I would be happy and fulfilled being a postdoc for the next two to six years. I thought not. But of course being a postdoc was only a temporary position. The real goal was to work your way up to associate professor, eventually becoming a full, tenured professor at a place like Princeton University. So I looked around, thought about all the tenured professors I worked with, and realized that I did not want to be one. I have nothing against physics, but it was clear that academic life was not in my blood. Given the realization that another 15 years of work was itself not something I desired, then it was crazy to continue on that path. If attaining a position similar to that held by your boss's boss is not your goal, then you should seriously consider whether you're in the right job now.

Things moved quickly from there. I called back the recruiter, restarted our year-old dialogue, interviewed at a large hedge fund in Hoboken, New Jersey, was offered a job in their research group, and took the position. To be fair, I was also exploring other opportunities ranging from Microsoft in Seattle to NASA's Jet Propulsion Lab in Pasadena. But none of them panned out, and my heart was really set on the hedge fund opportunity anyway. I left Princeton in March of 1993 without the particular support of my friends or faculty. From their point of view, I was abandoning the cause. My family by contrast, and especially my wife, provided all the encouragement I needed, and that made a world of difference. My experiment was complete, but I still had some writing to finish and needed to spend every commuting hour on the train (of which there were many) finalizing my thesis document. At last I defended my research in the spring of 1994, was granted my PhD, and never looked back.

And that, quite simply, is how I became a quant.

A Circuitous Route to Wall Street

Of course, it wasn't really that simple. In the days leading up to the hedge fund interview, my recruiter insisted on coaching me about my clothing (a new suit was mandatory), how to answer questions (come "prepared" with detailed ideas), and even what to eat (salad or pasta, little or no sauce, and make sure to always use the knife). I knew nothing about finance, nothing about the markets, and even less about hedge funds. But in order to be "prepared," I gathered my thoughts on how the markets might behave if they could be modeled as physical processes. It struck me that although market prices must be affected by supply and demand pressures, these forces are themselves affected by changes in prices; this type of self-referential feedback reminded me of a particular quantum mechanics process called renormalization. I got it in my head that if you could strip out all the supply and demand forces from an observed price using quantum renormalization techniques, then you could arrive at a true fair price. This type of thinking was sure to make a damn good impression at my interview.

The hedge fund was located on the Hoboken waterfront, which required me to pass through some seedy parts of Jersey City on my drive up from Princeton. I got lost. I was late. I arrived out-of-breath, with my new suit all disheveled. The office was posh, and the receptionist offered me a mint while I waited for my interviewer. After 15 minutes I was greeted by a man . . . dressed in jeans and wearing a leather jacket. Not at all what my recruiter told me to expect. He asked if I was hungry and decided we should head out to a noisy local pub for some chow. I just hoped they had no-dressing salad on the menu.

The conversation started after we ordered. He asked about my background, and in particular what I might know about finance. Here then was my chance to impress him with the 'Berman Theory of Market Renormalization." I began by telling him about my work in physics, which he very much appreciated; so much so that he asked me if I was one of those "nuts" who thought you could apply quantum mechanical principles to the financial markets. I'd like to be dramatic and say that time stood still after he made that comment, but in reality only an instant passed after which I said "of course not, it's crazy to think you can apply the theories of quantum mechanics to financial markets." We laughed, had dessert, and I got asked back for a second interview.

This time we went out for pizza. We each ordered two slices and a couple of Cokes. The pizza came out dripping with grease and I noticed he grabbed a stack of napkins and blotted his slices to soak up most of the oil. I had grown up eating New-York style pizza my whole life and never saw anyone "blot" a slice. So, I grabbed my own handful of napkins and proceeded to do the same, at which point he exclaimed, "Wow, I never saw anyone else do that before – I thought I was the only one!" I got the job.

My work during that next year was incredibly rewarding. The focus of the fund was to create automated trading strategies and apply them to global futures markets, including commodities, equities, and fixed income. As long as it was a valid futures market, we traded it, regardless if the prices represented Eurodollar contracts or Red Azuki Beans. I spent a lot of time writing very complex code to create and backtest different types of trading strategies using daily futures data back to the 1940s. Oodles of data, challenging analyses, and lots of programming—this is exactly what I had been doing in physics for a dozen years, and I was groovin'.

But alas, I quickly came to realize that finance is not rocket science. After all, I was a rocket scientist and I knew the difference. This is because in physics statistical distributions arise from fundamental physical processes that can usually be modeled, and therefore future distributions can be predicted with amazing accuracy. This is not the case in finance and very rarely does anyone try to model an underlying fundamental process. Instead most everything starts with, and ends with, observed distributions. Not only do I not believe quantum mechanics can be applied to financial markets, but I think most of the advanced statistical analyses financial folks perform today are ill-applied to the point of absurdity. There is a big difference between astronomy and astrology, but not everyone realizes that. And that's what made me very good at this first job. While spending many a long lab night staring at printouts from spectrum analyzers and traces from particle detectors, I routinely observed hundreds of six-sigma events appearing and disappearing right in front of my eyes. In order to succeed in experimental physics, you had to build up a real intuition about what types of results were real, and what was nothing more than data-mining, curve-fitting, or just plain statistical anomaly.

By the end of my second year of research at the fund, I had gone about as far as I could go; but fortunately so had one of the fund's general partners with whom I happened (not accidentally) to have developed a very good working relationship. He had decided to branch off and create a personal family-and-friends fund that would combine the existing systematic trading strategies we were using with an overlay of fundamental stock and commodity analyses. He had all the capital he needed and asked me to join him in his new venture. Ever the opportunist, I agreed.

We crossed the Hudson and setup shop under the auspices of ED&F Man in the World Financial Center, right in the heart of downtown New York City. As a two-person operation, my first task was simple—recreate, from scratch, everything that the previous 30-person fund had done, but in a way that could be wholly automated and required no additional staff. Over that next year I coded day and night, and even purchased a $20,000 Sun SparcStation laptop (that's right, a laptop) so I could code during my two-hour-per-day train commute. I was in heaven. I created my own futures backtesting language, a byte-code compiler, and an automated web-based trading system. With these tools I developed many new styles of trend following that had never been done before at the previous fund. Each night the system would upload the latest closing prices for each futures market, rerun my simulation routines, generate signals, and auto-fax trades to our London brokers for execution the next morning. Everything worked flawlessly except during market holidays when rogue prices would trigger bad trades. Much to my wife's amusement, I had to keep that laptop in our apartment bedroom to make sure I could unwind those false signals when they arose.

After a year or so, we hired a second person and started the next phase of our strategy: fundamental stock and commodity analyses. And it is here that I really learned about the markets. Rather than just looking at streams of prices, I spent time studying the cattle cycle, reading crop reports from Iowa, and tracking silver shipments from overseas vaults to COMEX warehouses. While my system continued to trade in the background, we placed targeted option bets whenever we felt a particular commodity was about to come into severely short supply.

By spring of the second year, our fund was up almost 40 percent, and I informed my wife that my anticipated bonus would be large enough to provide us a nice down payment on a small house. But by the end of summer we had given back over 30 percent, and poof, our new-house dreams had vanished. My boss was unperturbed as he had already been quite wealthy and was trading more for the game than for the money. But one of the directors of ED&F Man was not happy and he made sure to let me know it. Man had very little money in the fund, so the give-back of profits was inconsequential. Nevertheless, this director wanted to know how I, as an ex-graduate student previously making $15,000 per year, could so easily let real money go like that. Wasn't this a significant amount of money to my wife and me? Shouldn't I have taken some profits? After a stern lecture he parted with the following advice: "Once a month you should do what I do. Get right up close to a solid wood door, and . . ." BAM! He appeared to smack his head squarely into my office door. "There, that's the only way to knock some sense back into yourself!" And with that he walked out.

Over the next year and half, I spent a lot of time studying trend-following systems and examining the commodities markets. My goal was to fully understand why our strategies sometimes worked and sometimes did not; I therefore found myself keenly focusing on the fundamentals of risk management. Eventually, I came to some solid conclusions about trend following and convinced myself these strategies would no longer produce the same level of profits in the years ahead than they had in the years behind. I also found that that I had less of a stomach for the huge ups-and-downs of trading then I had previously believed. This was a hard lesson for me to acknowledge. After all, many people claim that if they just had some money to start with they could use it to make a fortune. Well, I had millions in potential capital at my discretion and was asked to do just that. But I couldn't. I very much wanted to, but I just couldn't come up with any systems or strategies that would make a fortune in the commodities markets. The more I wracked my brain, the more I realized I liked the detailed analytical parts of my job much more than the trading parts. It seemed to me that making money in the markets involved a much higher percentage of luck than was compatible with my academic persona. To rationalize this, I conjured up the image of money being like snow—sometimes it just drifted in a huge pile to cover one person's car, while the car parked next to it had hardly a flake. The only thing it meant was that one car happened to be upwind for this particular storm and the other was not. But I needed something where working harder and becoming smarter proved to be an advantage rather than just a distraction.

The Last Mile

By the end of my third year at ED&F Man, I called back my original recruiter and asked him to start a search for a financial position that best suited my analytical (mixed in with some real-world business) background. A week later someone from his office phoned to ask if I were interested in a possible position at JP Morgan. Or at least I thought he was from my recruiter's office. It turned out that he was actually from a completely different recruiting agency. I asked him how he knew I was looking to switch jobs and he said he didn't, it was just a cold call. But how did he even get my number? Easy, one of the human resource folks at JP Morgan was an ex-Princeton undergraduate and asked this recruiter to search out ex-Princeton graduate students in a hard science like physics or chemistry to fill a potential role in equity research. This recruiter got hold of a Princeton alumni directory and started calling physics graduates. Since my last name began with a "B" I was at the top of the list. Note to readers: always keep your contact information up-to-date in alumni directories!

I jumped at the chance to interview with JP Morgan and a few weeks later found myself in their Broad Street offices. After some small talk with the ex-Princeton HR person who had started the process she asked me what I knew about Morgan and of equity research. I admitted I did not know much since most of my hedge fund experience pertained to commodities research and trading systems, but I did admire a document from JP Morgan I came across that previous year. And that was the RiskMetrics Technical Document published in 1996. It was one of the finest pieces of true research I had seen come out of Wall Street, and it fit nicely into my academic view of how research should be done.

She smiled and asked if I knew anything about RiskMetrics other than what I read in the Technical Document. I said I did not, and with that she reached for the phone and made a quick call. Again she smiled and asked if I could keep a secret. It just so happened that the JP Morgan group responsible for the RiskMetrics methodology was in discussion to spin out of Morgan and create a new standalone business dedicated to risk management research and analytics.

I found it very hard to sit in my chair. A few more phone calls, and within 10 minutes I was on the 19th floor of 60 Wall Street interviewing with the folks at soon-to-be RiskMetrics Group. For me, this was an ideal venture, as it combined many elements of my academic background with a strong entrepreneurial twist that nicely echoed my past business experiences. After a few days and a dozen or more intensive interviews I was finally offered the job, and joined RiskMetrics in September of 1998 on the day the group spun out of JP Morgan. I had traveled that last mile and was now, quite literally, on Wall Street.

It's been more than eight years, and I'm still at RiskMetrics. I've held quite a number of positions, arranging from product manager (where I started), to heading the market-risk business, and to even running the sales force. I've learned an enormous amount these past years, about myself, about Wall Street, about finance, and about the many types of people who work in this sometimes strange, sometimes wonderful, but never dull industry. Am I still a quant? Of course, but now I apply my quantitative skills just as much to negotiating business deals and setting up organizational structures as to modeling hedged fund risk and scaling a parallel-processing architecture. As I said at the start, being a quant is not a job, but an attitude, and one that I hope will serve you as well in your career as it has in mine.

Chapter 4

Evan Schulman

Chairman, Upstream Technologies, LLC

To measure is to know.
If you cannot measure it, you cannot improve it.

—Lord Kelvin

This epigraph is inscribed over an archway at the University of Chicago. Passing under that arch one day, Frank Knight, the noted economist, is reputed to have remarked, "Then let us go forth and measure everything."

One of the first of my own measurements was marking my students' final exams. I had taken a job teaching introductory economics at the University of Western Ontario. Measurement confirmed that I had little effect on the students. In economic jargon, my marginal productivity was zero or worse. I knew at that point I was in the wrong profession.

The Royal Trust Company in Montreal gave me my first job in the real world. I became a junior trust officer. Duties there were mainly administrative functions to the widows and children of those deceased, fortunate enough to have acquired sufficient assets during their lives. A few months into this career, I approached the head of the department and suggested that there must be a better way to exploit whatever talents I had.

Showing rather remarkable sensitivity for a trust company back in the 1960s, the firm gave me a budget at the McGill University computer center to explore the potential application of computers to financial markets. I set out immediately to measure things, and by so "knowing" I would be able to suggest improvements.

Measurement

One of the first exercises was to measure the performance of pension accounts administered by the trust company. These accounts were valued monthly or quarterly, as needed: cash flows, in and out of the accounts, occurred randomly between valuations. What rate of return had been earned by these investments over various different time periods and asset categories? Clearly this was an appropriate application of Lord Kelvin's need to measure, if we were to improve the process.

Making the heroic assumption that valuation changes could be prorated evenly over the time between valuations, I used a binary search routine[1] to find the rate of return that generated the observed ending market values given the recorded cash flows and beginning values. We could now measure and report to the firm's clients an approximation of the dollar-weighted investment performance of their funds. The trouble with dollar-weighted performance was that it is client specific; it depends on the actual cash flows of the client whose performance was being measured. To generalize performance measurements the industry developed time-weighted performance. This was calculated from unit values, similar to the reported performance of mutual funds.

Pension and endowment clients, much like the clients of mutual funds, learned to use time-weighted performance reports to "buy" performance. Unfortunately time-weighted past performance was a highly visible, but unreliable, indicator of future performance. I am not certain that these first steps in measurement were a help to the investment community or to the most efficient allocation of invested capital.

Market Cycles

It was clear that clients lost money when markets fell and, equally apparent, the trust company lost revenue. Could computer intensive statistical techniques extract signals of future performance from observed prices? Many months (and computer cycles) later, I learned there was a dramatic difference in the accuracy of predictions using within-sample data and those tested on out-of-sample data. I was well on the way to becoming a convert to the efficient market hypothesis.

Inaccurate predictions were not just the result of overt data mining. Because of the work I was doing, I was allowed to attend investment meetings between the trust company and some of its most prestigious corporate clients. One meeting in particular comes to mind. I listened with rapt attention to authoritative dictums and insightful analysis. The pension fund had raised its cash reserves a few months prior due to clearly observed economic uncertainties. However, the economic clouds had since dissipated, and the market recovered nicely. It was time to reduce the reserves to normal working levels to avoid the rather large opportunity costs the fund was experiencing.

The reader will have correctly anticipated that within a few months of implementing this decision, stormy economic clouds returned; the market reacted unfavorably, and the pension fund lost real money on top of the opportunity costs mentioned above.

Although this was only one observation concerning a high-powered investment committee, it, and others like it, helped to tighten the distribution of my Bayesian prior (based on the previously mentioned in-sample, out-of-sample statistical work I had done at the McGill computer center) concerning efficient markets.

Process

Well, if we could not exploit market inefficiencies, could we improve the investment process? An opportunity appeared to look at how the trust company ran its own money. The trust company's treasury issued paper with fixed maturities based on its credit rating and invested the proceeds in the paper of other firms. Earnings were a function of spread in quality ratings and maturity extension. The regulatory authorities laid down a lattice of investment restrictions in an attempt to ensure that financial firms, such as the trust company, would be solvent and able to redeem their paper in a timely manner. The trust company met these restrictions with a set of manual heuristics, ensuring compliance by adding cushions to most, if not all, such restrictions. Could a computer allocate the funds in a more efficient manner?

This seemed to be an ideal opportunity for a linear programming application: Maximize expected return while meeting all regulatory restrictions. The results demonstrated that manual adherence to each restriction with cushions was expensive in terms of opportunity costs.

The appropriate matching of liabilities with asset classes was also a process worth exploring. The trust company had an airline pension fund as a client. A significant investment in that portfolio was real estate in Florida. In and of itself, this appeared to be a very risky investment. I had the temerity to point this out to the airline's pension officer. He noted that a significant number of the airline's pilots tended to retire in Florida, and that his job was to ensure stability of their retirement package, not in Canadian dollars, but versus the basket of goods they would consume upon retirement. I wandered down the hall somewhat confused. The argument clearly made sense as stated, but the airline had a defined benefit plan; as such, the retirement package was the direct liability of the corporate sponsor. Retirees would be paid according to their length of service and final pay levels, not according to the value of the pension assets. Further, not all their pilots retired to Florida. Not only had I swallowed the efficient market doctrine, but it looked like I would also be a proponent of defined contribution plans.

Risk

About this time, I received an offer to join the Keystone Group in Boston. From the point of view of a young immigrant, the investment department was staffed by a remarkable collection of talented individuals.[2] One dared not let the mind drift during investment meetings.

Keystone's hallmark was that it offered a set of risk-controlled stock and bond funds spanning the spectrum from conservative to speculative. Control was achieved by restricting the fund managers to select securities from specified buckets with little or no overlap. Performance was measured by how the fund did against the performance of the bucket. Altering the securities in the buckets was a relatively painless way to alter the relative performance of a fund. The process of maintaining the buckets was run by one of the fund managers. Could a quant improve on this process?

It was relatively simple to get the portfolio managers to agree on a set of securities that epitomized the risk buckets and the variables (price variability, size, P/E, etc.) that described such classifications. We then employed discriminant analysis to fill the buckets. This solution had some of the previous system's operational problems. For instance, how often should the system be refreshed because, with new data, securities on the margin could change buckets? And how should portfolio holdings be handled, given that the underlying securities were moved to a different risk bucket? But by and large, the portfolio managers now concentrated on running their portfolios rather than trying to change the measuring rod.

And Return

In terms of performance, the question of significance was raised. On the one hand, it was clear that the portfolio manager who was restricted to investing in the conservative group of securities had difficulty generating results that were dramatically different than the average of the bucket. On the other hand, managers picking from the more speculative universes often strayed dramatically, one side or other, from the average performance generated by the securities in that bucket. There was also the question as to whether the average performance should be calculated as an equal weighted or capitalization weighted average.

As luck would have it, Kalman Cohen and Bruce Fitch published an article titled "Average Investment Performance Index" in 1966 that described a closed-form solution that gives the average and standard deviation of returns of a set of randomly generated portfolios from a specified universe of securities.[3] The AIPI algorithm allowed users to specify ranges for the number of securities in the portfolios as well as for minimum and maximum security weights. Portfolio managers could thus be rewarded as a function of how significant, in statistical terms, their performance difference was. Given that, prior to this innovation portfolio managers received bonuses for overperformance, regardless of whether it was statistically significant, I was not altogether popular among my associates that year.

A last observation on return: High-risk portfolios fascinated me. The adage that return compensates those who accept risk produced visions of wealth. What would have happened had clients, or I, invested in the very-high-risk products offered and held that investment for the long term? Keystone had returns on their risk-controlled products going back 30 or more years. It mattered little where one started or ended the process: the longer the time period, the greater the shortfall between the arithmetic average performance of the high-risk funds and their compound returns over that time period. Harry Markowitz notes that the relationship between average performance and compound return can be approximated as follows: Compound returns equal a function of the average return less half of the variance.[4] Increasing the variance of returns without increasing the average return sufficiently leaves the investor poorer for the experience.

Trading Costs

There was no question that our investment department was talented.[5] The funds held securities that performed well. However, the performance of some of the funds, especially some of the more volatile funds, was disturbingly bad. What was happening? The periods involved were far too short for this to be the result of the compound return problem.

Whoever designed the reporting system at Keystone had the foresight to record the decision price. That is, the price was recorded at the time the portfolio manager decided to buy or sell a security. It was a relatively trivial but revealing exercise to compare the decision price to subsequent prices: It was clear that the fund manager had information; it was just that the fund manager did not have sufficient information to cover the transaction costs. Put another way, the market reacted quickly to news and backed away from aggressive traders, from concern that the trader had new information. The effects of trading costs are eloquently described by Andre Perold. "The Implementation Shortfall: Paper vs. Reality" article published in the Spring 1988 issue of the *Journal of Portfolio Management*.

Informationless Trades

In early 1974, the investment group at Keystone wanted to take advantage of the coming negotiated commission rates to rebalance its high-risk portfolio. But we were mindful of the market impact costs of such large transactions. Were there ways that we could signal other traders that for this particular trade, or set of trades, we were not trying to exploit proprietary information? We chose to divulge to a few broker/dealers the characteristics of the set of desired trades. We told them the number of stocks involved in both the buy and sell sides, the size of the positions relative to the average daily volume, the relative riskiness of the stocks, and the overall buy and sell packages. We told them everything but the names of the stocks. We then asked them to give us a fixed cost to execute the trades. The cost was to be measured from the closing price of the stocks on the night before we divulged the names of the securities to the executing broker.[6]

Michael Bloomberg, who was then the head equity trader at Salomon Brothers, understood what we were trying to do and made the most aggressive bid. Not as aggressive as I wanted, but clearly better than what we had measured in the past. In early May 1974, Salomon and Keystone executed the first program trade to occur after the advent of negotiated commissions.

Applying it All

Shortly thereafter, I moved to Batterymarch. The goal was to apply the lessons learned to date. We used the computer to apply quality control (discipline) to the investment process by ensuring diversification, controlling trading costs, and searching for sectors of the market that might be priced inefficiently. Small stocks seemed the most likely to be mispriced because they were not followed, valued, and traded by the institutional investors of that time. We screened large commercially available databases looking for measures of value.[7] Performance, as measured by alpha, was significant. The firm prospered, doubling assets under management every year for seven years. We also went international, noting that not all markets marched to the same drummer. This lack of correlation helped diversify client portfolios.

At Batterymarch, we worked very hard to lower commissions and keep the market impact of our trades to a minimum. With regard to commissions, we developed a computer system that brokers would log into and get orders that were plainly advertised thereon. We "sunshined" our orders to indicate that we were not information traders. Our advantage was the rigorous valuation analysis of publically available data. As a result, the contra side had little to fear in the short term. To further show we were not trading on short-term information, our orders would have a couple of substitutes. When an order was executed, the other orders were canceled. Traders who tried to front-run such orders could face the equivalent of an air pocket. We also posted the low commission rates that our clients would pay; brokers could choose to do business with us or not. Those who did so were ranked by how much business they did on our system. Every quarter we dropped the one or two who did the least business with us on the grounds that we were only an insignificant portion of their business and that we should give their places to other firms who might try harder. We continued to use blind bids for large restructuring programs.

Unfortunately, we were inadequately prepared for the day when small stocks ended their performance run. We, the principals, indulged ourselves with some finger pointing. It was time to move on.

However, while I was at there, we did establish the Batterymarch Fellowship as a token of our appreciation for the many contributions from the academic quants. This award was to fully support untenured member(s) of faculty. The award was paid directly to the named winner(s), bypassing the assessed overhead of the universities. We cast the net wide in an attempt to find undiscovered talent in the lesser known schools and, each year, sought a senior academic and a representative from our clients to help us evaluate the proposals and letters of recommendation submitted.[8] What we discovered was a relatively efficient market for academic talent. All the winners came from top schools. As a result, submissions dwindled from 30 or so at the start of the program to 15 or so later on as candidates self selected.

Electronic Trading

After the advent of negotiated rates, commissions fell over time from the 1 percent level to something like 0.15 percent of the value of a transaction. Volumes on the exchanges soared. Before negotiated commissions, 20 million shares traded was a big day on the New York Stock Exchange (NYSE). One-billion-share days are not uncommon some 30 years later. To compound the problem, the average trade size plummeted. There was no way this volume of transactions could be done manually. The exchanges and Nasdaq opened electronic access to their "floors." These paths were to help brokers deal with the plethora of small orders, leaving them free to concentrate on the large institutional orders.

Lattice Trading

It was clear that the volume of orders arriving at the institution's desk were also rising, significantly. Again, this was not a problem that could be solved by throwing more bodies into the mix. Lattice Trading,[9] first operational in 1992, was to give institutions, as opposed to brokers, electronic access to the floor. It was a multibroker system to allow for the fact that institutions tended to use several brokers in the conduct of their business. Lattice was designed to monitor prices, volume, and cash positions and to change orders and execute trades as a result of changes in these variables. For instance, limit prices could be hooked to the price of the S&P futures or any other security on the tape, order size could be increased or decreased in real-time according to a client's cash position, and partial orders were placed on an exchange with the remaining order hidden, but available to cross with a contra order on the Lattice system. Our dream was that institutional traders would place all orders on Lattice and then monitor their progress. The trader would only interfere when progress was insufficient.

Lattice operated in the *continuous markets* of the time—Nasdaq and the NYSE—it was even hooked to the London Stock Exchange. I liked to refer to it as a semi-intelligent trader. But it faced all the limitations inherent in continuous markets—namely, when you have a couple of million shares to buy or sell, do you really want to trade with the contra party who will satisfy your whole position? Don't you wonder what she knows that you don't? Also, the regulations were changed so that stocks traded, not with spreads of an eighth of a dollar or more, but with penny spreads or less. It was now easy to front-run a large order. Simply place an order a penny or less in front of that order. If the decision-making behind the large order was correct, you would profit from any executions you made. If the price didn't move in the right direction soon enough, you simply dumped your front-running executions into that large limit order below you for a penny loss. Limit orders became a fool's game.

The result is that brokers now offer institutions dozens of *lattices*. These are systems that take institutional orders directly to the floor, cutting them up into little pieces to avoid market impact. Clients get the volume-weighted average price, or the results from other mechanical execution strategies. Block trades became few and far between.

Net Exchange

During this time I encountered Net Exchange, a spin-out from Caltech with a much more elegant solution to the institutional trading problem.[10] Net Exchange was designed as a combined value call market-handling conditional orders that used AND, OR, and IF THIS operators. Users could enter orders to, say, sell one million shares of GM *and* buy 3.5 million shares of Ford *or* 350,000 shares of Caterpillar Inc. *and* generate $2.5 million in cash for this account. Further, if the volume is insufficient, prorate this order. Notice, there need not be a single contra party for this order. In fact, it is highly unlikely there would be. The orders filling this may come from a host of other users. Thus, users can place large orders, expecting that those orders will be filled by many contras, a number of which should be *noise* or liquidity traders. Also, the fact that users must wait until the end of the day for the "call" means that traders trying to take advantage of information that quickly becomes available to the market are less likely to participate: If the information becomes available before the call, the trader has no advantage.

What is also interesting is that price limits are not part of this order. The price limit here is the cash that the user wishes to extract from, or spend on, the trade. As long as these cash limits are at all reasonable, this type of order gives the computer degrees of freedom to enable it to execute trades. As a result, combined value trading makes the most efficient use of whatever liquidity is present at the call.

Peter Bossaerts at Caltech tried to validate the Capital Asset Pricing Model (CAPM) using MBA students at Stanford and Yale as traders. The environment met all the CAPM assumptions: All participants had same time horizon and the same knowledge of the securities. The market contained a riskless security, the security return distributions were normal and all the available assets were tradable and held by the investors, and so on. Bossaerts expected the students to trade their portfolios to attain the highest Sharpe ratio, which he had set to be that of the capitalization weighted market. He assumed the students would drive their portfolios to the zero line, as shown in Figure 4.1.

However, the results were all over the place when they traded using continuous market mechanisms like those used by the NYSE, Nasdaq, and London's SETS (Stock Exchange Electronic Transfer Service). Indeed, the students, who were well versed in finance, left potentially profitable arbitrage opportunities on the table because, when trading one security for another, they could not be certain of filling the second part of the trade at an advantageous price. The result of only partially completing the trade might leave their portfolios further from equilibrium than their current position.

Using conditional orders in a call market format (conditions AND, OR, and IF THEN), relatively few traders, read thin markets, were able to drive their portfolios to equilibrium very quickly. Call markets that allow conditional orders make the best of existing liquidity and are ideal for thin markets. I note that because of their size, institutions habitually face thin markets.

Conventional vs. Combined Value Trading

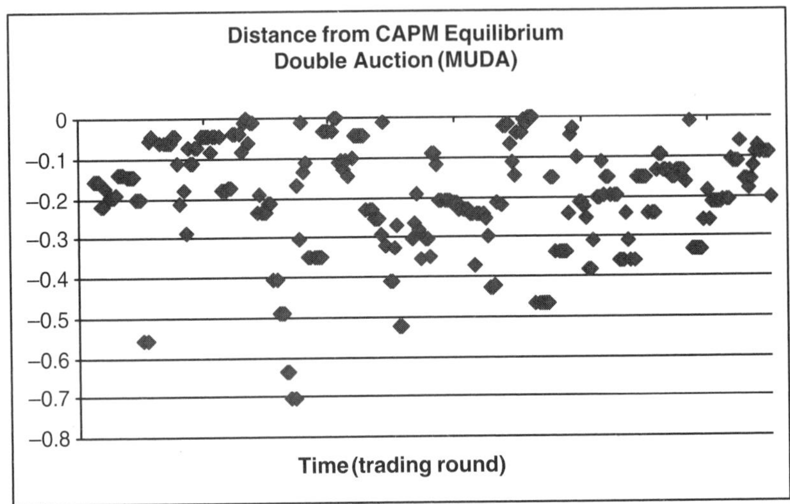

40 participants

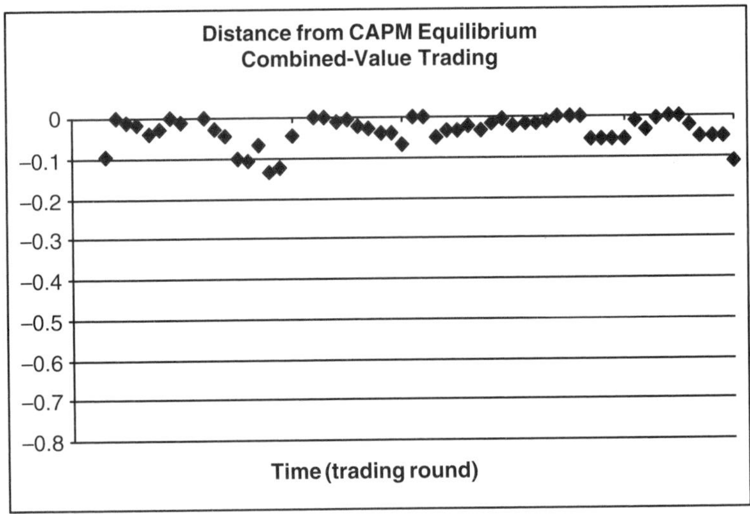

12 participants

Figure 4.1 Conventional vs. combined value trading.
SOURCE: Upstream Technologies.

Upstream

Lattice was sold to State Street Bank, and placed in its brokerage division. I learned once again that I was not suited for work in a large organization. Mark Hoffman and I moved on to establish Upstream Technologies in late 1999. The goal of Upstream was to apply the tools, discipline, and quality control that quants had developed for institutional use to individual accounts, even the very small ones. We recruited Paul Samuelson, previously chief investment officer at Panagora, and a set of talented youngsters.

It was clear that for the taxable investor individually managed accounts dominate mutual funds. Mutual funds generate taxable distributions (often including short-term capital gains), and investors are unable to take advantage of the unrealized losses in the fund's underlying portfolio. Properly constructed individually managed accounts offer the same level of pretax returns as mutual funds but can take advantage of short-term losses and avoid short-term gains, client by client. Further, they meet the current demand of investors for customized portfolios that take account of the beneficial owner's specific legacy holdings and investment restrictions.

The market is huge. The resulting growth in individually managed accounts at trust companies, registered investment advisors, and broker-sponsored WRAP accounts has been dramatic. Given the economic forces driving it, this growth will continue. However, quality control of the product has now become an issue, given the size and growth of the market and the complexity of the service offered. As with trading years before, it is no longer possible to meet demand by throwing more people at the problem. This would cost the manager too much. It is necessary then to find a way to scale the business such that clients get the customized service, which is what is driving the growth of this market, while allowing managers to gain efficiencies to make this segment profitable.

Technology has advanced so that it is possible, using standard servers, to optimize individual portfolios at a rate of something like two per second, and the application can scale almost linearly as users add hardware. Optimization is ideal for the accounts of individuals. Using optimization, centered on a selected lead account, model or benchmark, managers can control the tracking error versus the lead or benchmark, sculpting portfolios around investment restrictions, legacy holdings, and tax problems. Models and security rankings can be imported from any source and managers can set up multidisciplined portfolios using either composite models or asset sleeves (miniportfolios). Trust companies, registered investment advisors, broker representatives acting as portfolio managers and others, who act as portfolio managers, are potential users.

The application of this integrated tool set of model management and portfolio management, when combined with an order management system, allows money managers to significantly alter their production function. Since the administrative duties of customizing client portfolios are now expertly and consistently performed by machine, management firms can significantly reduce their spending on junior portfolio managers and errors and increase it on client services, sales, marketing, and getting the content (the models or lead accounts) right. It is possible to assure clients that the manager has processes in place to evaluate and analyze *their* specific portfolio, given *their* specific tax and investment problems, every day and, if the market presents opportunities, the manager's investment process will exploit them in the client's interest.Further, such a system addresses two problems currently plaguing the industry, fiduciary oversight and the capacity to deliver product.With regard to oversight, the system can log the reasons for each and every trade (who, why, how, stock ranking, marginal contribution to risk, tax consequences, transaction costs, etc.) as the portfolio manager sees them when the trade is authorized. Those charged with oversight should make it clear that they will question all trades generated manually rather than by the approved processes of optimization or replication. Manual trades should become few and far between, allowing those with oversight responsibilities to concentrate on getting the investment content of the lead accounts or model portfolios to correctly reflect the investment insights of the firm. This form of discipline produces quality control and permits the firm to profitably scale its operations.

Some money managers close their mutual funds, other investment products or strategies to new business because trades in the securities used for the product/strategy are seen as having undue impact on the market. Optimization that focuses on tracking error can produce funds that track a model or benchmark without using the particular securities in that model or benchmark. Generating the same product with different securities mitigates the capacity problem. For those who argue that it is the particular securities, not the factor exposure that generates the sought-for returns, I suggest ranking three or more securities as investment candidates in each investment sector considered. Wait a couple of months and correlate those rankings against the performance observed over those months. Do this for a number of sectors for a number of time periods and you will develop both a sense of humility and an appreciation of the random walk hypotheses.

Articles

During my travels, I have coauthored three articles, the essence of which is not included in the stories above.

In 1998, Andre Perold and I coauthored "The Free Lunch in Currency Hedging: Implications for Investment Policy and Performance Standards."[11] We argued that, because currency boasts a long-term expected return that is close to zero, the sizable effects of currency risk can be removed with minimal transaction costs without the portfolio suffering much of a reduction in long-term return. Fischer Black immediately followed this in 1999 with his classic "Universal Hedging; Optimizing Currency Risk and Reward in International Equity Portfolios."[12]

In a 1999 paper with Ross Miller, "Money Illusion Revisited: Linking Inflation to Asset Return Correlations,"[13] we extended the pioneering work in behavioral finance by Franco Modigliani and Richard Cohn. They showed that investors tend to undervalue firms in inflationary times if they do not properly account for the effects of inflation on a company's income statement. They termed this effect *money illusion.* Our paper examined a corollary of their result: In the presence of money illusion, the correlation between stock and bond returns will be abnormally high during periods of high inflation. For the United States, it was shown that inflation had exactly this effect on stock/bond correlations during the postwar era. As a result, asset allocation strategies that are based on the high correlation coefficients calculated using data from the 1970s and early 1980s can be expected to generate inefficient portfolios in regimes of low inflation.

Ray LeClair and I wrote a paper, "Revenue Recognition Certificates: A New Security," in which we explored the concept and potential benefits of a new type of security.[14] This security provides returns as a specified function of a firm's sales or gross revenues over a defined period of time, say 10 years, and then expires worthless. To answer the question as to whether certificates would advantage our markets, we discussed the benefits to investors and issuers and reviewed some of the agency problems associated with conventional debt. We suggested that certificates may be a cheaper financing solution for firms than debt. This should be true if any one or combination of the following hold: The price of the implied put associated with certificates is less than the implied put in debt financing and/or management assigns costs to the restrictive covenants associated with debt instruments and/or investors see portfolios of certificates as a natural hedge against unexpected inflation or value the security more because it is more transparent with regard to the underlying return generating process. We also thought certificates might be more liquid because of the transparency issue and the fact that its investment characteristics change dramatically as it approaches maturity.

These articles and the discussion on electronic markets do raise questions: If we were given a blank sheet of paper and asked to design a way to finance firms that would also be a reasonable basis to fund insurance and retirement claims, would we come up with bonds, stocks, defined benefit plans, and continuous markets? As Fischer noted:[15]

"More effective capital *is* more capital."

Chapter 5

Leslie Rahl

President, Capital Market Risk Advisors

I was born in and grew up in Manhattan, the financial center of the universe (at least according to Saul Steinberg's drawing that appeared on the cover of *The New Yorker* in March of 1976). Yet Wall Street didn't figure in my thinking during my youth. The road I took, while it eventually led to the Street, was not a straight line.

Growing Up in Manhattan

My Manhattan roots are deep. Both my parents were born and still live in Manhattan. My dad was a dentist and his office was on Wall Street. That was my only real connection with Wall Street. He had originally intended to be a medical doctor, but was locked out by the Jewish quota system used by medical schools of the time—and he was told his grades were too high! It was, possibly, portentous for me that during the interval in which he was deciding on his second choice career, he took a master's degree in statistics.

My mom had a been music major in college and still, at 83, sings three times a year at Carnegie Hall with the Collegiate Chorale. She taught music in a nursery school and became a stay-at-home mom when I was born. Later she became involved with teaching the deaf and received her master's degree from Teachers College at Columbia. Thus, while both parents were college educated, neither had a quantitative career. We lived on the Lower East Side on Grand Street, a working-class, largely Jewish/Italian neighborhood, and when I was two and a half, I was enrolled in the Downtown Community School (DCS), a small, private school on 11th Street, where I remained until eighth grade.

Somehow I had acquired math and reading skills before I started school, so my teachers, to keep me productively employed, taught me to knit and then allowed me to practice knitting and to attend second-grade math while my classmates went through their math and reading exercises. Fortunately for me, though, perhaps, a loss to the fashion industry, I was allowed to skip the second grade.

The school was "progressive," so I can recollect no one discouraging me as a girl from excelling at and enjoying math. *Progressive* is perhaps not the correct description for the DCS philosophy. I think DCS was run by card-carrying members of the Communist party, so that I was regularly lectured on the evils of capitalism in general and of the United States in particular. This political aspect of my early education, which neither my parents nor I grasped until eighth grade, I believe actually prepared me very well for my eventual college experience.

I enjoyed math throughout elementary school. It seems a bit eerie thinking back on it, but I was always able to readily assimilate new math concepts as they were presented to me with matter-of-factness (a skill that eventually eluded me, when I was introduced to differential equations). At age 13 I volunteered to tutor high school students in math, and also helped a girl at the Henry Street Settlement to prepare for the math SAT and get into a community college.

After I was graduated from the DCS, I attended Friends Seminary for high school. At Friends, I again found myself enjoying math and science more than English and history. My high school math teacher,

Mr. Lannon, used to say, "The impossible we do right away; the more difficult takes a little longer." His teaching and encouragement helped lead me on my path be a quant. More than 35 years later, I dedicated my book *Hedge Fund Risk Transparency: Unraveling the Complex and Controversial Debate* to him. It became apparent that I would work in a quantitative field, the particular field at the time still undetermined—and, as it turned out, not yet invented! As a high school senior, I took the math SAT IIs (or whatever the achievement tests were called back then) with eight of my classmates (there were only 25 in my graduating class), and seven of us received 800s in math.

College and Graduate School

I chose MIT for college, in part because I was concerned that a female math major might be seen as weird at other schools, while I felt pretty sure that at MIT I would be considered normal. Even so, when I arrived in Cambridge in the fall of 1967 with the rest of the class of 1971, I was one of about 50 women out of 1,000 freshmen. MIT had enrolled coeds since 1861; but before 1966 they had had to be from the Boston area, as all freshmen were required to live at home or in a dorm, and there was no women's dorm.

I loved everything about MIT, and, in retrospect, it was clearly the right choice for me. In my first semester, I enrolled in 6.251 (everything is referred to by a number there), "Compilers, Loaders, and Assemblers," the introductory computer science course that preceded the establishment of a computer science program at MIT. Since the class was part of Course 6 (the Electrical Engineering Department), I assumed that I had discovered my calling and set about pursuing the goal of becoming an electrical engineer.

Today 6.251 is called Introduction to Mathematical Programming, still part of the Electrical Engineering degree, but with not even a hint of the prior incarnation of the course. My electrical engineering aspirations have likewise long since disappeared, and I rely on others to plug in equipment and change light bulbs.

I followed 6.251 with classes in computer linguistics and programming (which I loved), but then found that I neither liked nor had any aptitude for circuit boards, electricity, or the mechanical facets of the electrical engineer's trade. Thus disillusioned, I was delighted to learn that MIT allowed students to petition for a customized course of study. I petitioned to combine classes from courses 6, 15 (management), and 18 (math), to create my own precursor to a computer science degree. I had the privilege of programming on the TX-0. The "tixo" (Transistorized Experimental Computer Zero) had 8K of core memory. I also worked on the Arpanet, the precursor to the Internet. I was a whiz at "punched cards" and was employed the summer after my sophomore year to program the "Print Using" function for GE's BASIC language.

Those were still the days when the slide rule ruled, and I remember, consistent with my dislike for mechanical things, hating to use a slide rule. However, I was very proficient with what we called the *CRC Blue Book Tables,* in reality, *Mathematical Tables from Handbook of Chemistry & Physics,* by Hodgman, Selby and Weast. This book was published by the Chemical Rubber Publishing Co., now known as the CRC Press, a transformation no doubt motivated by the same type of thinking that has more recently given us KFC as a successor to Kentucky Fried Chicken.

The course 15 (management) classes I took as part of my custom program of study corresponded to the first-year requirements for a Sloan degree, so I decided to stay at MIT for a fifth year to complete the MS program. MIT did not offer an MBA at the time, only an SM (master of science). Today, both are offered, but those who write a thesis get an SM; those who don't get an MBA. To prove that even at MIT the ugly head of bureaucracy rears itself on occasion, I had to petition Sloan to waive my first-year requirements, which I had already taken and passed as an undergrad, else I would have been required to take the same classes a second time.

In the year I was completing my SM work, the first preference-free option pricing model was being developed and discussed in the business school at MIT— completely unbeknown to me, as I was concentrating in the subfield of decision theory. My master's thesis was developing a scoring model-based algorithm for screening and selecting interns for Tufts New England Medical Center. My algorithm, based on simple rules, generated a very accurate replication of historical selections.

The social scene during my first year at MIT in 1967 included a dance, complete with gowns and long gloves, and parietal hours (when students of the opposite sex were allowed in dorm rooms as long as the door was kept open). Starting in 1968, my sophomore year, at MIT as throughout academia, the atmosphere became very politically charged. Yet the campus protests seemed to me very civil. When the students would strike, the faculty would strike along with them. The administration had agreed with the Cambridge police that university security alone would handle the protests, unless the University requested Cambridge police assistance.

One of my suite-mates (the Crown Princess of Thailand) was very politically active. When she participated in a protest, it was always strange to see her surrounded by the protective shield of her personal bodyguards. To me, the political situation of the day didn't seem too dark, at least in comparison to the picture that had been painted by my elementary school teachers at the DCS, who inadvertently helped me to stay out of extreme currents of political activity and stay focused on my education. The United States couldn't be as bad as I had been taught at DCS!

When I look back at my postcollege job search, I am amazed by the haphazard process and the factors that came into play. I had no idea what I wanted to do or where I wanted to work. I limited my consideration to firms that recruited on campus. I used none of the techniques that I now regularly advise others to use. I didn't create lists of pros and cons; I didn't visit a company again after receiving an offer. I went with my gut.

While interviewing with Texas Instruments, I learned that there were no women in management positions at that time. All women there were apparently employed in clerical positions. Several of my interviewers asked me who I thought I might sit with in the cafeteria (apparently, they were concerned that I would defect from management and sit with the secretaries). Taken aback by this, I told my interlocutors that I assumed I would sit with whoever I became friends with. I also told Texas Instruments no thanks when they offered me a job.

Ultimately, I chose to enter Citibank's management training program, rather than work at any of the technology firms with whom I had interviewed. My attraction to Citi and finance had less to do with my love of finance or math than it did with being impressed that Citi chose to inform me of their offer by sending me a telegram—the first telegram I had ever received. My first year's salary was to be $15,500, a vast sum in 1972.

It was still a year before Black and Scholes published their seminal option pricing paper (though, as a result of the perversity of the academic paper review process, the empirical part of their research appeared in print in 1972), and a year before listed equity options trading would begin.

Nineteen Years at Citibank

When I joined Citi, the bank had had only two women VPs in the worldwide organization. They were not allowed to eat in the VP dining room. Even several years later, after the bar had been removed, the downtown VP dining room still had no women's bathroom.

I started as an officer in the corporate bank. I didn't want to become a loan officer, however, and was able rather quickly to move out of the management training program into the first of several staff management positions. These positions were staff jobs (financial control, strategy, public relations, and finally, chief of staff of a large division of the bank), not quant jobs. In fact, I was finding that my most useful class from business school was the one that had been, at the time, the most derided—a required course in managing people.

After five years passed in this way, I moved to the Operating Group, where I ran large, operationally intensive businesses. The first one I ran supported the banking needs of Wall Street—a stock vault, daylight overdrafts, and so on. Subsequently, I ran the Stock Transfer and Corporate Trust Division for two years, and then Global Investing (i.e., ADRs [American Depository Receipts] and precious metals). One motivation for my switch from staff position to operational business was an attempt to move out from staff roles more common for women into more of a line role. I was not completely successful in this because, even though at one point I had a large P&L and 250 staff, my businesses did not have the cachet among senior managers of "line" P&L generating businesses.

Therefore, I was ready for another move when the head of the investment bank asked me to become his chief of staff late in 1981. I was reluctant to return to a staff role, but accepted, subject to the quid pro quo that after a year I would hire a staff, and organize the function, he would teach me how to trade. I knew almost nothing about trading. When, early on in my staff prioritizing, I was asked by my new boss to prepare the division's quarterly review, I was at a complete loss as to what to write for the section titled "Trading Strategy." As a placeholder, I wrote "Buy low, sell high." The head of the investment bank loved it, and the report went directly to John Reed as written.

I had not advanced far in my trading education when fate intervened and gave me a big push. In 1983, the Chicago Board of Trade (CBOT) began exchange trading of options on bond futures. I was offered the opportunity to trade these derivatives as a proprietary trader for the bank's account because "you don't know anything about trading, but at least you went to MIT. Maybe you can figure the option things out."

I was responsible for building and trading a proprietary options book. The market was not efficient and the technological firepower of desktop computers was still embryonic (and we didn't have any yet). We used a Monroe calculator to price options (although we knew it was incorrect).

The trading floor was not only male dominated, but also had a culture based on machismo. Fortunately, my experience at male-dominated MIT helped prepare me for life on the trading floor! I am not sure what the Cambridge post office had thought when every month at MIT I received eight copies of *Playboy* magazine, subscribed to on behalf of male friends who were afraid it would be stolen from their dorm/fraternity mailboxes.

As an extension of the proprietary options book I was trading, I was able to build an interest-rate caps business. If the men on the trading floor had had any idea that this activity would evolve from a small P&L source to a major business, or be as "sexy" as the M&A business, then I would have not been given the chance. We hired an academic to build a pricing model. The model provided Black-Scholes type prices, but with a couple of simplifications. To be able to obtain real-time prices at the beginning of each day, the head trader (me) had to select the values to assign to each of two parameters. The first parameter was yield-curve shape, and the choices of parameter values were "relatively flat" and "relatively steep." The second parameter was spot volatility (no vol surface here!). The choices were "relatively high" and "relatively low." This model was more sophisticated at the time than models used elsewhere in the bank and among our competitors for cap pricing.

The cap business was client driven. One of the early clients came to the bank looking for a long-term cap, but didn't want to pay a significant premium (other things equal, longer-term options cost more). To accommodate the client, we "invented" the interest-rate collar. A *collar* combines an interest-rate cap with an interest-rate floor, and the floor offsets part of the cost of the cap. The client was so satisfied with the idea that we agreed to a deal for $100 million notional over ten years. I smile when I think that we dynamically hedged the risk of this position with a cash bond. Still, in the early days of this market, all trades were bespoke, and the correspondingly high profit provided a significant cushion against hedging errors.

Eventually, the swaps business was added to my responsibility for the interest rate options business. In fact, it was having this combined portfolio of exposures that allowed me to survive the crash of 1987. At the time, the options book was massively structurally short, while the swaps book was very long the market, so we ended slightly in the black for the week. Postcrash, I was determined to never again be structurally short or long anything! My quest for a source of long-dated volatility led to the invention of the *swaption*.

Fifteen Years (So Far!) Running Capital Market Risk Advisors

My trading experience has given me a healthy respect for model risk. In many cases model-based prices are best used as indications of whether to buy or sell, and *not* for valuation. I am a big fan of marking-to-market whenever possible. I think people get mesmerized by the sophistication of their models and the numbers they generate. Maybe two-thirds of the risk management tool box is qualitative, not quantitative tools. It is essential for a risk manager to be able to combine judgment, market savvy, and common sense with quantitative skills and tools.

In 1990, I had a baby, left Citi, and started my own consulting firm. In 1994, that firm became Capital Markets Risk Advisors, with derivatives as a core competency, but an escalating focus over the years on the buy side and on hedge funds.

I have had the opportunity to take on an extraordinary range of assignments in the past 15 years, for 275 clients on six continents. We undertook the investigation of Bankers Trust ordered by U.S. financial regulators after the derivatives losses experienced by P&G and Gibson Greetings. I consulted to the Fed after the LTCM crisis.

I was called by Orange County a month to the day before its bankruptcy and asked "to review their portfolio and analyze what the impact of the margin calls they were starting to get would have on their yield." We were the ones to have to explain to them (and the SEC) that they had more serious problems than a decrease in their yield! We arranged four bailout plans during their final weekend meltdown, but they were not able to enter into any of them. Once Citron was fired, no one at Orange County was legally authorized to enter into a deal.

In many ways, the most satisfying assignments I have had over the years have been those that have help clients develop approaches to grow and thrive. The strategic use of risk is a favorite assignment. Despite my preference for constructive, forward-looking projects, I do view my experience as an expert in several high-profile litigations (Askin, Sumitomo, MinMetals, etc.) as unique opportunities for gaining insight and wisdom that I regularly share with clients as to what NOT to do.

Running my own business for 15 years has been the highlight of my career. I love having no one to blame but myself when things don't go the way I'd like—and the freedom to take on the assignments that intrigue me and decline those that don't. I also love the flexibility of not having to worry about "face time" with the boss and being able to work when it suits me.

There are, of course, significant differences between running your own business and working for a big company. The first time the phone system went out I was painfully reminded of the level of support that those in the corporate world enjoy. On net, the positives far outweigh the negatives. Running my own business has also provided me the freedom to attend schools plays, chaperone school trips, and attend parent-teacher conferences. Running my own business has also allowed me the opportunity to take on both for-profit and not-for profit board assignments (or, as a friend of mine from England calls it, *going plural*).

Going Plural

It has also enabled me to make a significant commitment to philanthropy and community service. I started a mentoring program with other MIT alumni to help middle school students compete in Lego Robotics tournaments. I was on the board of and chaired the philanthropy committee of "100 Women in Hedge Funds" from 1991 to 1994 and helped raise $5 million. I started "High Water Women" in 2005 to encourage other senior women in financial services to give back, and we raised over $1 million our first year, fulfilled "Secret Santa" wishes for 600 homeless children, and bought new backpacks to kick off the school year for 750 children in need. In February 2004, I joined the Fannie Mae board and have had a wild but exciting and fulfilling ride. I also joined the MIT Investment company board ($11 billion) and the NYS Common Retirement Board ($120 billion). I guess there's truth in the adage, "If you need something done—ask a busy woman."

The Personal Side

I met my first husband on the first day of freshman orientation at MIT and we were married in the middle of our junior years and stayed together for 10 years. I used a PERT chart to lay out the organization of our first dinner party—I must be a true quant. I met my second husband, the love of my life, at a wine tasting, and we have been married for 17 years. By the way, I took wine tasting for Independent Activity Period (IAP) at MIT. In addition to our wonderful 16-year-old son, Kevin, I have a stepdaughter Kaitlin, who was seven when she first starting living with us half-time and is now engaged, and a stepson Stephen, who was four when he started living with us half-time and who is now 23 and works for a hedge fund.

So How Did I Become a Quant?

So how did I become a quant? I'm not sure you *become* a quant—I think I was *born* a quant. To me a quant is someone whose eyes don't glaze over when they see a formula. Someone for whom numbers speak as loudly as words. Someone who is extremely logical. (My husband doesn't always agree that I possess this component of quantdom—but I think I do!) Being a quant is *not* the same thing as being a nerd (although they are sometimes confused). To be a nerd requires white socks and a pocket protector. I'm proud to be a quant, and believe that not only does it help me with quantitative problems, but has taught me an analytic framework for problem solving that applies equally well to nonquantitative problems.

Chapter 6

Thomas C. Wilson

Chief Insurance Risk Officer, ING Group

I am honored to be included in this book among many of the industry's leading financial engineers, I must admit, however, that I accepted initially with some trepidation and somewhat under the feeling of false pretenses.

You see, I am not sure that I consider myself a quant. Throughout my undergraduate studies in business at UC-Berkeley, I took the absolute minimum mathematics and statistics courses required—so few, in fact, that I had to undertake an extensive crash course before I began my PhD at Stanford! I have never found a mathematical proof *intuitive* or *elegant* or even vaguely inspiring, but rather, more often than not, a slightly painful exercise. And, although I am reluctant to admit it in public, whenever I think of Martingales, I am just as likely to think of small songbirds pleasantly singing on Sunday mornings as statistical theory.

Yet, paradoxically, I find myself truly honored today by the label of *quant*. Naturally, this has been the cause of some serious introspection in light of my perceived lack of aptitude in quantitative areas. Receiving this honor has caused me to question why I have apparently used mathematical finance and statistics consistently throughout my career as a risk professional. Did quantitative methods play the lead or supporting role and, if supporting, what exactly was the lead?

As I try to summarize the source of my contributions to the industry, I believe that it lies in the early phrasing of *the question* rather than in the application of any specific quantitative technique, no matter how useful the tool may be for answering the question. Fortunately for me, there was no lack of important and challenging questions raised by the rapid developments in the risk-management world over the past 20 years, especially in the area of market risk in the early 1990s, credit risk in the late 1990s, and, now more than ever, shareholder value creation.

Over my career, I have contributed to each of these areas; at times, I have had to learn new statistical, mathematical and financial techniques. As a consequence, my path to becoming a quant has to be seen in the context of the industry developments that I describe in the rest of this section. I then conclude with some advice to the aspiring quant, including outlining some of the lessons that I have learned during the course of my career.

Quantitative Finance: The Means to an End?

In trying to characterize the role that quantitative methods have played during the course of my professional career, I first considered, and then discarded, several different metaphors. Each of these metaphors invariably related to the use of quantitative techniques as tools, or as the means to a specific end. While these metaphors may be useful to describe the path of some, they do not describe the path I have taken to (apparently) becoming a quant.

One metaphor that I discarded was that of a craftsman's tools: just as a master woodworker requires a fine plane, clamp, and calipers in order to shape an elegant violin from a block of raw wood, so too does the financial engineer need quantitative finance tools to develop, for example, elegant option pricing techniques for new and innovative structures and underlying processes. Alas, I cannot claim to have produced, like Stradivarius, an elegant option pricing formula out of a simple block of wood.

A similar metaphor considered, but also rejected, was along the lines of the more mundane builder's tools: just as a framing carpenter uses an air hammer to efficiently frame a wooden house, and quickly gets something structurally sound and robust but perhaps not as fine as the finishing carpenter's work, so to does a financial engineer use numerical techniques such as finite differences, binomial trees, variance reduction techniques, and Monte Carlo simulation to make the theoretical, practical. Alas, I also cannot claim to be a builder of sound numerical foundations on which the theory can be implemented.

Neither of these metaphors seemed appropriate in my case. Both required a degree of expertise that I cannot claim, and both focused on the tools as opposed to the end itself. In fact, quantitative techniques as tools did not play the lead role in my contributions but rather a supporting role. What, then, played the "lead" role? As I was thinking about this, I was reminded of an observation I made while at Stanford University completing my PhD At the time, I thought that it was possible to divide the finance and economics academic community between two broad camps along a single continuum.

I put on one end of the continuum those individuals who have used financial mathematics or numerical techniques to solve well-established and practical questions. If pressed, I would describe such renowned individuals as Duffy, Hull, Morton, and Jarrow as having made contributions to theory, using financial mathematics to develop option-pricing formulae much as Stradivarius must have used plane and calipers to construct his violins. I remember reading the work of Heath, Jarrow, and Morton with reverence in the 1990s as they first established the existence of a broad class of arbitrage-free term-structure processes and then went on to characterize them. Although many might understand what the end product should look like, few are capable of wielding the tools to create the violin, or its financial equivalent. Also at this end of the continuum, I would put such individuals as Singleton, Sargent, Engels, and Granger for their contributions on empirical and numerical approaches used to make the theory practical. It is their work, building the foundation in terms of empirical parameterization and numerical solution techniques, that ultimately makes the theory practical.

I put on the other end of the continuum individuals whose contributions were driven more by "the" question, or the intuitive interpretation of the observed economic and financial phenomena, rather than by the quantitative techniques that were used to represent their intuition. For individuals at this end of the spectrum, phrasing the question seemed more important than the techniques used to find the answer. In this camp, I put such individuals as Akerlof, Stiglitz, Lucas, Diamond, and Dybvig, individuals whom I judge to have contributed more through the intuition behind the question then the actual quantitative techniques they used. Who can argue that the intuition and insights behind Akerlof's market for lemons outshadows the relative simplicity of the algebra used to prove the point?

Although I do not claim to be the peer of the individuals in either camp, I like to think that the use of quantitative techniques during the course of my career more closely resembles the latter, intuition-based camp than the former. I have always enjoyed identifying and framing the question earlier than others, and been conscious that answering the question, once phrased, would nonetheless require me to learn new quantitative tools.

The Questions

Fortunately, important, interesting, and challenging questions with substantial business and managerial implications have presented themselves during the past 20 years. They fall roughly into the three main areas of market risk, credit risk, and shareholder value creation. Although many of the questions and answers in each of these areas are taken for granted today, they often challenged conventional wisdom at the time.

In the remainder of this section, I summarize the historical context and the questions that I found interesting in each of these areas. For brevity's sake, I have excluded some themes, including my contributions into asset/liability management for both banks and insurance companies.

The Early 1990s: The Market Risk Era

The early 1990s marked the *market risk era*. During this period, the banking industry developed economic capital or value-at-risk (VaR) models, Raroc performance measures,[1] and treasury funds transfer pricing rules in order to answer questions with substantial strategic consequences. In order to understand these developments, as well as the contributions I made during this period, it is useful to provide some historical context. Although complex and difficult to summarize, several forces were at work in the early 1990s that helped to shape the market risk and treasury agenda. Primary among these was "The Great Derivatives Debate."

In the late 1980s and early 1990s, the derivatives market grew at triple-digit rates, from nothing to a multitrillion-dollar industry. In 1990, Gerry Corrigan, then president of the New York Federal Reserve Bank, echoed other central bankers' concerns over this rapid development and the possible systemic risk involved. He called derivatives "financial time bombs" wielded by quantitative college grads. Based on these concerns, he issued a stern warning to the industry to improve the level of risk management in the industry.

As a consequence, a new and innovative regulatory framework was developed by the Basel Committee on Banking Supervision with substantial input from the banking industry. It covered market risk through the use of internal VaR models as well as new derivative credit exposure rules based on current and potential future exposure. Without the development of VaR models for the market risk of the trading desk and potential exposure models for the credit risk of derivatives, it is doubtful that the derivatives industry would have been allowed to continue to develop at the same accelerated pace.

These developments opened up new fields in applying quantitative approaches to the measurement of risk, both in terms of the theory as well as in finding practical numerical solutions. Within this context, the questions that I chose to ask, and answer, were central:

- *Is it possible to calculate VaR for nonlinear portfolios from local risk information (e.g., deltas and gammas), which are readily available on the trading floor?* If the industry could answer this question, it could calculate

VaR more accurately and cost effectively for its nonlinear trading portfolios. I believe that I was the first in the industry to answer this question with the delta-gamma approach (see Wilson (1994b)), based on the observation that a quadratic form of normal variables is distributed as the sum of noncentral chi-squared variables for which numerical solutions are available.

- *How many independent factors are practically required to capture the risk of a multicurrency fixed income trading book?* The dimensionality of VaR calculations for a global trading book quickly becomes too large to be calculated efficiently, especially if each point on the yield curve is modeled separately. A logical place to look for a reduction in the dimensionality was therefore in the modeling of multicurrency interest rates. My answer (see Wilson (1994a)) compared both factor analysis and eigenvalue decompositions of multicurrency term structures and attempted to characterize the required number of factors and the stability of the parameter estimates over time.
- *What happens to the tails of our VaR calculations if we have only estimates of volatilities and correlations and not their exact values?* Many in the industry at one time treated volatilities and correlations as known when, in fact, they were only estimates made with uncertainty, and therefore introduced parameter risk in their own right. I believe that I was the first in the industry to illustrate this effect in Wilson (1993) using subordinated processes and showed that the estimated market risk distribution would have fatter tails if the uncertainty surrounding volatility and correlation estimates were taken into account.

During the early 1990s, as the industry was coming to grips with the great derivatives debate, I identified these as interesting and important questions very early, often before others. Unfortunately, I did not have the tools at the time to answer them directly. Instead, I had to learn new quantitative techniques in the areas of subordinated processes, factor decomposition, and multivariate distribution theory. I became a quant not because I applied tools I knew, but because I had important questions that could only be answered using tools that I did not know at the time.

The Late 1990s: The Credit Risk Era

As the industry got comfortable with the measurement of market risk in the mid-1990s, it naturally turned to credit risk, leading to the second important period in the development of risk measurement techniques. At the time, the credit risk measurement framework at the individual name level was well established, including the decomposition of credit risk at the obligor level between the probability of default, exposure at default, and loss given default. However, the analysis at the portfolio level was not as advanced. Was it possible to develop a portfolio credit VaR model, similar to those being created to measure market risk? What were the key questions that needed to be answered in order to do so?

Two of the most important approaches during this period included the actuarial-based model, CreditRisk+, from CSFP, and the modified-Merton model as implemented by KMV. This was a period devoted to the theory of credit portfolio risk measurement; unfortunately, few institutions stepped back and asked whether the theory matched the empirical data at the portfolio level. Although much work had been done documenting some models' credit default predictive capabilities at the individual obligor level, there had been relatively little empirical investigation of whether models fit aggregate portfolio behavior.

In 1997, I developed a credit portfolio model, sometimes called the Wilson or McKinsey model and officially named CreditPortfolioView™ after the software implementation, which was distributed freely. Using this model, I raised some fundamental questions and challenged conventional wisdom in many areas:

- Do equity markets alone provide sufficient information to characterize the contemporaneous correlations between average sector default rates and migrations?
- How useful are other factors such as the level of real economic behavior as measured by GDP, interest rates as a surrogate for debt coverage, or foreign exchange rates for sectors with international competitors in explaining sector defaults and their correlations?
- Do equity markets, which react instantaneously to new information, effectively describe the autocorrelated time series properties of credit cycles?

- How can retail portfolios and large corporate exposures be integrated into the same portfolio model consistently?
- How can default only portfolio models be expanded to a mark-to-market approach comparable to market VaR?

In order to address these questions and develop a model of credit portfolio risk covering both retail and large corporate exposures, I had to learn many new techniques in the area of statistical theory, econometrics, and numerical solutions. Although the new techniques were interesting to learn, my primary motivation was to answer the challenging questions that were only partially being addressed by the industry at the time.

The Great Strategy Debate: From the 1990s to Today

In parallel with the development of the derivatives market in the early 1990s, many traditional commercial banks were faced with a fundamental strategic issue: After watching the corporate loan market become commoditized due to intense competition and disintermediation in the capital markets, many commercial banks seriously considered changing their strategic focus. Some leaders such as Bankers Trust and JP Morgan underwent a fundamental transformation from a commercial bank to trading institutions. As with any transformation of this magnitude, the process was difficult and required strong commitment by senior management.

Raroc models, used to make the risk-adjusted return from different businesses or loans directly comparable, were developed by these leading institutions in an effort to sharpen their strategies and build the commitment of senior management. In retrospect, the development of Raroc and its strategic impact have been self-reinforcing, leading the industry to look differently at the economics of credit businesses and the measurement of value creation within a financial services company. Without the development of Raroc models, it is doubtful institutions would have had the courage and the commitment to fundamentally change their corporate credit strategy and practices.

This trend in using Raroc measures to guide corporate strategy has continued throughout the 1990s and into the next decade as institutions have reinforced their focus on shareholder value creation. For example, most financial services institutions use Raroc as a cornerstone for their Economic Profit or Economic Value Added™ framework, guiding their investment in and development of different lines of business. As a consequence, the identification and correction of any possible bias in Raroc has become even more important today than ever.

These developments have also opened up new fields in applying quantitative approaches to the measurement of performance, both theoretical and empirical. Within this context, I raised several questions that were central to the debate. For example:

- *What are the inherent theoretical biases, and is it possible to make adjustments to correct for the biases?* In the early 1990s, the entire banking industry was moving headlong toward Raroc as a pricing and performance measurement framework. However, as early as 1992, I recognized that the common Raroc measure based on own portfolio risk or VaR was at odds with equilibrium and arbitrage pricing theory (see Wilson (1992)). Using classical finance to make the point, I recast a simple CAPM model into a Raroc performance metric and showed that Raroc based on own portfolio risk without the recognition of funding was inherently biased. In the years since 1992, many other authors have followed a similar line of thought.
- *What is the appropriate cost of capital, by line of business, if capital is allocated based on the standalone risk of each underlying business? And, what role does earnings volatility play in the valuation of a bank or insurance company?* Many in the industry hold to the belief that an institution can use a common hurdle rate across different businesses if each is capitalized to a common rating standard on a standalone basis. As I had shown in 1992, economic theory, including the simplistic CAPM model as well as the later work of other authors, suggested otherwise. More importantly, simple comparisons of the cost of capital for similarly rated investment banks, retail banks and insurance companies indicated that shareholders require greater returns from companies that have more systemic risk, even after normalizing for their capitalization and risk of default. In 2003, I undertook an

empirical investigation based on a series of nested models to prove this point (see Wilson (2003)).

Although most financial institutions currently use Raroc, its use is becoming more nuanced as the link between Raroc, the debtholder's perspective and shareholder value creation continues to be challenged and improved. My early contributions in this area were theoretical and, later, empirical, forcing me to undertake a whole new line of econometric investigation and hopefully adding clarity to the debate.

Lessons Learned

Based on the path that I took becoming a quant, if I had to give advice to the aspiring quant today, it would be to get *the* question that is relevant for today's world "right." You can always learn the quantitative techniques you might need to answer the question if it is phrased appropriately—and is important enough!

In addition to this advice, I would also comment that quantitative techniques are a useful and occasionally a necessary tool for financial engineers to answer important strategic and tactical pricing questions. As with any tool, however, they can also be used inappropriately. (I am reminded of an old adage, "To the man with a hammer, every problem looks like a nail.") During the course of my career, I have learned several valuable lessons on the use of quantitative techniques, which I summarize here briefly.

- *Build your intuition before building your model.* It can be frustrating for a model builder to start a model in the hopes that it will ultimately tell him something interesting, much as it must be frustrating for a magician to try to pull a rabbit out of any old hat that just happens to be lying around. Here is the secret of every successful magician: The rabbit is already in the hat before the show starts! The moral of this fable: Before you start to build a complex model, make sure you know what results you want and intuitively where your rabbit is, or what feature of the model will generate the results that you want. If you don't build the rabbit in, you will most likely not find your rabbit!

- *Trust your intuition.* Sometimes, complex models can provide unanticipated results that challenge our intuition. More often than not, the unanticipated results are caused by something simple such as an error, a misspecified regression equation, and so on. Before throwing out the old theory, check your algebra and trust your intuition!
- *Challenge your intuition.* Very often when developing complex models, it is easy to get lost in the layer upon layer of commonly used assumptions and rational, which lead one to accept results unchallenged. Consider two of my favorite hobby horses: Why should it be the case that shareholders will ask the same expected return from a single-A investment bank as from a single-A property and casualty insurer just because they are capitalized to the same probability of default? Or, why should all information on the future development of credit default rates be solely encompassed in equity prices? Always challenge your intuition and the status quo.
- *Be as good at communication as you are at the theory.* Some are good at modeling, others are good at explaining, but few are good at both. If you want your models to have significant strategic or practical impact, it is useful to be able to explain them in terms mere mortals can understand, without resorting to the Greek alphabet or stochastic calculus.
- *The model is always wrong—but that doesn't make it useless!* The beauty of a mathematical, self-contained system is that, based on the assumptions, it is always right. The reality is that your assumptions are always wrong or, at best, an abstraction; as a consequence, your risk or pricing models will be wrong with 100 percent probability. That doesn't make them worthless, however! It just means that you should always use good judgment when using a model, and make sure that it is appropriate for the circumstances in which it is to be applied.

Chapter 7

Neil Chriss

Former Managing Director of Quantitative Strategies, SAC Capital Management, LLC

I started my finance career when I left a postdoctoral position in mathematics at Harvard University and took a quantitative research position at Morgan Stanley Institutional Equities Division in New York City. When I moved to Wall Street, by all accounts my academic career was going fine. I had the good fortune of having a terrific academic position and was actively pursuing a number of interesting research projects. The reason I left academia had to do with the types of problems I enjoyed solving and the question: Do I want to devote my life to solving pure math problems?

Many aspects of life are different when viewed from the outside than when view from within. For research mathematics, this is particularly true. As much as it may seem like getting a PhD in mathematics and then doing research at Harvard would indicate a serious devotion to the subject, the truth is that, compared to many mathematicians, my level of interest and devotion paled. I liked it and I worked hard at it, but it was never clear to me that I could spend my life doing it. One of my problems was that I didn't care about the big picture.

In my first year of graduate school, I read a biography of David Hilbert, one of the great early twentieth century mathematicians. There was an anecdote that Hilbert was asked about the first thing he would do if he awoke one day and found himself transported 500 years into the future. Hilbert's response was that he would find the nearest mathematics department and ask if the Riemann Hypothesis—an important and difficult problem that remains unsolved to this day—had been solved. To me this is the level of interest in mathematics you need to do it your whole life. You have to care deeply about the subject.

So then why do mathematics at all? One answer is that it is fun to work on solving math problems and if you can do it for a living, why not? Hilbert himself commented on the importance of mathematics problems and noted in a famous address to the International Congress of Mathematicians in 1900, "a mathematical problem should be difficult in order to entice us, yet not completely inaccessible, lest it mock at our efforts. It should be to us a guide post on the mazy paths to hidden truths, and ultimately a reminder of our pleasure in the successful solution." In other words, we work on problems that are pleasurable to solve but they must also be possible to solve—hard, but not too hard—and they should tell us something deep about mathematics. Great problems are pleasurable to solve in part because they are connected to the larger scope of Mathematics—math with a capital "M." For me, the question was did I care enough to do this for the rest of my life or was there something better for me? All of this is reminiscent of a famous novel, Hermann Hesse's *The Glass Bead Game.*

The Glass Bead Game

So why *not* do mathematics? One reason is because it is so disconnected with the rest of the world. *The Glass Bead Game* is a favorite novel among my mathematician friends. It takes place in the future at a time and place in which a monastic order of intellectuals has split apart from society. They engage in a complicated, all consuming game—the Glass Bead Game—and strive to achieve mastery of it. The novel is fascinating because it never really describes the game in full, but rather alludes to a game based on mastering relationships between seemingly unrelated things, as in this quote:

> A Game, for example, might start from a given astronomical configuration, or from the actual theme of a Bach fugue, or from a sentence out of Leibniz or the Upanishads, and from this theme, depending on the intentions and talents of the player, it could either further explore and elaborate the initial motif or else enrich its expressiveness by allusions to kindred concepts. Beginners learned how to establish parallels, by means of the Game's symbols, between a piece of classical music and the formula for some law of nature. Experts and Masters of the Game freely wove the initial theme into unlimited combinations.

To become a master of the game requires total devotion. To members of the order, the game is an all-consuming way of life. The game is abstract on the one hand but relates closely to the world on the other. Within the order its pursuit is paramount yet success in the game has no influence on the greater good of the world. Finally, the game has become so abstract it isn't even played with pieces any longer—it is a game solely played in the mind. Despite its complete disconnection with the rest of the world, to members of the order it represents everything that is important in the world. Each strives only to achieve the greatest honor of becoming *Magister Ludi*—Latin for Master of the Game. The novel is the life story of one Magister Ludi and his ultimate rejection of the order in order to become more connected with the "real world."

I think anyone who has done research in pure mathematics can relate to the *The Glass Bead Game.* Hearing its story, it is hard to believe that Hesse wasn't describing research mathematics. Despite how completely abstract and otherworldly pure mathematics is its pursuit seems all important when engaged with it. For example, here is the first sentence of my PhD thesis, "An unramified p-adic group is a linear algebraic group over a p-adic field k, which has a Borel subgroup defined over k, and which splits over an unramified extension of k." At the time I wrote that, it seemed perfectly reasonable that I had spent years putting that together. The point of course is that to me at the time and to people familiar with the area of mathematics I studied, this was a perfectly comprehensible sentence. In fact, to some it is a perfectly obvious statement.

Returning to my main point, another reason to consider *not* doing mathematics is that it is hard to do well and it is ultimately very competitive. From the outside, academia—where much of the research in pure mathematics takes place—may seem placid or even sleepy, but inside it is anything but. Mathematics may be isolated from the real world, but mathematicians are not isolated from one another. The field of mathematics has a natural hierarchy. Mathematicians generally work on research problems. There are problems and then there are hard problems. Mathematicians look to publish their works in journals. There are good journals and there are great journals. Mathematicians look to get academic jobs. There are good jobs and great jobs. Mathematicians want to do well *relative* to one another. It is hard to do mathematics and not care about what your standing is.

In Wall Street every year bonus numbers come out, promotions are made and people are reminded of where they stand. In mathematics, it was no different. In small and big ways, people were always jockeying for position. Whether it was the natural chatter about who got what job or who published in which journal or whose thesis problem was more important, it was definitely competitive.

Even in graduate school, I found that everyone was trying to see where they stood. When I was a first year graduate student at University of Chicago, two undergrads from a top undergraduate school came and stayed with me to check out the mathematics department. Both of these guys had been superstars in the Mathematics Olympiad competitions when they were in high school. This was an area of mathematical geekdom from high school that I had never been involved with and also had a bit of an inferiority complex about. I wasn't very good at the problems so I didn't bother trying.

As it happened, a friend of mine asked me to help her solve a particularly tough Olympiad problem. It was some complicated request involving her brother who was a senior in high school. All I remember was, if I solved, it there was a free lunch in it for me. My strategy was simple. I brought the problem to my houseguests and politely asked them to solve it for me. They said sure. But after a couple of days they told me they couldn't do it. I was struck by this because they were supposed to be the best. Well, given the situation and the fact that the pressure was off, I decided to take a stab at it. In general, I had no hope of solving these sorts of problems. I had read Richard Feynman's autobiography *Surely You Must be Joking Mr. Feynman,* and in the book he recounts how he had a reputation for solving really hard integrals just like the one I was saddled with. He claimed that his reputation came largely from knowing a certain trick that no one else knew.

I thought, wouldn't it be cool if this worked for this problem? Now that the pressure was off and it didn't matter whether I could solve it or not, I gave it a shot, more or less with the plan that I would try Feynman's trick and that was about it. Well, it worked, just like the book said it would. It was quite striking actually because there was nothing to it.[1] Happy with my success, I showed my houseguests, thinking they would be happy too. Their reaction was anything but. They sort of looked at each other, shook their heads, and told me that what I had done didn't make sense and didn't work. In short, they rejected my answer. A day later they came to me and said that they had taken another crack at the problem and had solved it using a different technique. Most importantly, my answer was wrong. I had $\pi/2$ and they computed $-\pi/2$.

Solving the problem was fun, and no doubt the competitiveness was all good natured. But it is also a case the problem taking on a different importance to all of us as our perceptions of it changed. At first I had no interest in solving it. But once I knew the guys had not solved it, I thought it would be fun to take a shot. For the guys, they clearly had only a passing interest in the problem until they had said they couldn't do it and I said I could. At this point they were able to find a solution and took the time to point out that my answer was off by a factor of -1.

Of Explorers and Mountain Climbers

Another possible answer to the question why *not* do mathematics is that life is short and there are lots of other interesting things to do. And, some of those things might be more fulfilling, more interesting, or more suitable.

When I was an academic mathematician it slowly dawned on me that I might not be cut out for a life in mathematics. For one didn't like the isolation from all things real world. Second, doing research mathematics has a certain character to it that I felt I was not ideally suited for. I used to tell people that great mathematicians are like the great explorers. Explorers want to discover something new—a new trade route, a new land, gold—by venturing into the unknown. They do so with no guarantees of success. Great explorers are characterized by their courage and their desire to discover something new.

Consider Christopher Columbus. His great passion was to find a westward route to the Indies. He doggedly pursued funding for his expedition, first from Portugal and finally from Spain, offering a new trade route in exchange for being named governor of any new lands he discovered and the title of "Great Admiral of the Ocean." After many attempts, Columbus secured financing from Ferdinand and Isabella of Spain and sailed off in 1492 to discover his westward route to the Indies. Instead, he discovered America.[2]

What does this have to do with research mathematics? The analogy is that to make great discoveries you have to take risk. You have to venture into the unknown. In fact, all of research mathematics is comprised of working on problems that have not been solved and by definition no one knows for sure if they can be solved. Some problems are harder than others, and to assess how hard a problem is we look at how long it has remained unsolved and how many great mathematicians have failed to solve it.

To my mind, the greatest mathematicians are like explorers—they search for solutions to problems that no one knows how to solve and they risk achieving nothing. A terrific example of this is the famous Fermat's Last Theorem.[3] Fermat stated his famous theorem around 1637 and great mathematicians of every generation tried to solve it until Andrew Wiles finally cracked it in the early 1990s after famously spending seven years working on the problem in secret and alone. Why did he do it? Why does any mathematician work on hard problems? Consider the stakes. In the case of Fermat's Last Theorem, no one knew for sure whether we could solve the problem at all. The fundamental question—"Can it be solved?"—had not been resolved. Yet Wiles risked seven years of his career on a hunch that he could solve it.

I cannot answer why anyone would take that risk, but I can say that to be capable of this represents a certain type of personality, one that I deeply admire, but one that is probably a little different than mine. Yet to me, to be able to do mathematics at the highest levels, one has to be capable of doing what Wiles did. In other words, it was not just that Wiles solved a hard problem. He also took a certain type of risk.

So if mathematicians are like explorers, and I am not an explorer, what else is there? I used to tell friends that I was more like a mountain climber. I liked to be able to see the summit before I begin my climb. Mountain climbing has plenty of risks—people die climbing mountains—but the risks are different. In mathematics research you cannot see the summit. In fact the summit may not even be there. On the other hand many other pursuits in life, challenging, rewarding pursuits, are much more like climbing to the top of the mountain. The challenge and excitement is in figuring out how to get there, not in whether or not *there* is even there.

Computers

Thinking back it is surprising that I did not to graduate school in computer science or into a high-tech job after college. As a junior high and high school kid, I was about as immersed in computers as anyone I have ever known. I learned to program when I was 11 years old on a TRS-80 home computer that was originally a gift for my brother.[4] Pretty much from the moment the computer arrived home, I dominated its use. I was absolutely fascinated with writing computer programs. I learned by taking the example program book provided with the computer and typing every program into it and then modifying each program little by little to learn how to do things myself. By the time I was 14, I had a Commodore VIC-20, an early color computer.[5]

Like many kids my age, I was obsessed with video games and preferred to spend as much time as possible playing them.[6] From the start, one of the reasons I was so attracted to programming my brother's computer was the idea that I would be able to write my own video games. As soon as I became adept enough in BASIC—the simple language that was installed in both the TRS-80 and the VIC-20—I started writing games.

At first my games were horribly. But slowly I got better at programming and was constantly attempting to make my games faster and more like their real-life counterparts.

Around the time I was 14, I decided to try and write a real game. I had this idea in my head that I could write a game and get a company to pay me for it. By this time I had taught myself assembly language, the native microprocessor's language, which allowed me to write programs that were significantly faster than in BASIC. Looking back, what strikes me is that I was much more interested in getting my name published than I was in making money. I think this frame of mind is one of the reasons I originally went into academia—I liked the idea of publishing my work. I actually sold two pieces of software for the VIC-20. The first was a graphics package that created BASIC commands for creating geometric objects such as lines and circles. The second was a videogame called D'Fuse.[7]

Writing D'Fuse was my first contact with the world of business. I wanted to produce a commercially successful game. That meant a game that was good enough to be shrink-wrapped and sold to tens of thousands of people and played without system crashes or bugs of any kind. To this end, I worked extremely hard at every detail of the game and engaged my friends to constantly come over and play the game so that I could find bugs and fix them. I think it took about six months from start to finish and I did most of the work my sophomore year of high school. It was my first experience with a long-term project that I saw through to completion from start to finish.

I wrote the game having no idea exactly how I was going to sell it but the whole time I had my eye on that prize. That said, I enjoyed writing it so much I figured if I didn't sell it, it would still have been well worth it. In any case, I was confident that my game was good enough that it would attract someone's interest. Once the game was finished, I took a straightforward approach to marketing. I looked in the various computer magazines I read and found the phone numbers of the companies that sold computer games. I would call up and ask for the person in charge of buying games (as if there were such a person) and simply told them about my game. I got mostly rejections. I didn't realize how strange my call must have seemed to many of these companies. At last, though, a small New Jersey company called Tymac took an interest.

I visited Tymac, demonstrated my game, and they agreed to license it from me.[8] We called the game *D'Fuse*[9] and the back of the package said "D'Fuse—By Neil Chriss," something I had insisted upon as part of the deal. I was extremely proud to have my own game with my name on it. In the end, the game didn't sell very well because of an unfortunate twist of fate. Just as my game reached the market, Commodore came out with a new computer, the Commodore 64, with 64K of memory and much better graphics. The VIC-20, and my game, quickly faded into the background. This never bothered me. For better or worse, I felt my job was done once I had sold the game. I quickly moved on to new ventures.

Writing D'Fuse was my first software venture. I enjoyed writing the game but found what I really liked was writing programs, so the next summer when I was offered a job writing programs for an educational testing company on the Apple II computer, I jumped at it. In this job, I had my own office, a printer, and disc drive—it's hard to believe now, but these were two things I did not have writing D'Fuse—and an interesting project to work on. The job was to write a system that would read in a text file and produce a test form that had some of the words removed. Students would take this test by filling in the words. My system had to prepare the test form, let students enter their answers, and then extract them for grading by a different system. This is all very basic stuff now, but at the time there was no cookbook answer of how to do it on an Apple II computer. The challenge was to make the system fast and easy to use while also impossible to break.

By the end of the summer, I had produced the testing software. This was all very gratifying. I had completed the task and delivered the product. Also, I had appreciated the freedom my boss gave me in letting me go about building the software more or less as I saw fit. All in all it was a great summer. I probably would have worked there every summer but for my less-than-stellar negotiating tactics with the president of the company right at the end of the summer. By all accounts, I believed I had had a successful summer. The programs I wrote were in good working order, but I thought I should be making more money. I was making $7.50 an hour, and I figured I should be making at least $15 an hour. The company president asked me if I would like to come back the next summer. I said yes, but I wanted to earn double my current rate.

I think this was really my first experience being yelled at by someone other than my parents. The meeting had started off with lots of compliments about the terrific work I had done and how they would love for me to continue working there. I thought this was just about right—in fact, I was very proud of my work and probably felt that they had no idea just how wonderful it was! Yet, when the word *double* came out of my mouth, the entire tone of the meeting changed. I don't remember what he literally said, but as I remember it, it came out something like, "Why you little . . . we have MIT graduates who don't make that much," That pretty much ended my career in educational software. Much to my parents' horror I stubbornly refused to negotiate and thus the job went away.

The next summer, through the help of my father, I got a job working for a consulting company that was doing work for Lederle Laboratories tracking clinical studies for their drug testing. As it turned out, this job paid much better and was a lot less interesting. In this job, I was given a system specification for a database and had to implement it in a language called DB2. There was very little room for creativity, I worked with the consultants, not the people at Lederle but at the same time was paid $300 per day for my efforts. I remember once, out of shear boredom, I created a test database of animal studies and populated all the animal names with things like "Adorable Puppies" and "Cute little kittens" because one of the consultants had told me that the labs were depressing—you could hear the animals crying and whimpering in the background. Somehow this database got mixed up with the real test database and ended up being presented to a senior team at Lederle, who no doubt had zero sense of humor about such things! My benefactors came to me the next day extremely upset and gave me a much needed lecture on maturity!

But what about the money I was making? I worked about 10 hours a day so my rate worked out to roughly double the double I had asked for at the educational company. How could I reconcile that one company wouldn't pay me $15 an hour for work that I considered to be creative and independent and much higher quality than the work I was able to bill $30 an hour for? The answer was obvious even then. I was not being paid for the work entirely in terms of how difficult it was to produce. This made sense to me even then. I knew I did not know what each company could afford to pay, how important my work was to them and what their other options were outside of working with me. My pay at both places very well could have been completely rational. The experience though had one bit of lasting impact on me. I became extremely skeptical concerning how good a measuring stick money is. This had ramifications for me when I left academia way down the line because I was cynical about going into things where the main measuring stick is money given my early experiences. But that is getting ahead of ourselves.

College Years

I went to University of Chicago for college and majored in mathematics. My choice of major can probably be best described as *not computer science.* After programming practically every day since I was 11 years old, I was pretty much sick of computers and at the time had no interest in pursuing anything related to them. In fact, I didn't even apply to Caltech or MIT, two schools that would have been natural choices given my high school experience.

For most of college, I did not touch a computer. By my senior year that changed. At the end of my junior year I audited a course in neural networks and my professor recommended me for a job working at Fermilab,[10] joining a research team that was trying to use neural networks to analyze nuclear collision data. My job would be to write a Fortran program to implement a neural network to find what are known as *b-quark jets* in reactor data. This was my fourth job using computers, but this time I would be working for research physicists who had an interesting problem to solve. I did not know it at the time, but this was the work I did in mathematics that was closest to what we do in quantitative finance.

My job was to work with my bosses—Myron Campbell from University of Chicago at the time and Bruce Denby from Fermilab—to figure out how to make a neural network discriminate between "interesting" collider events, those that would be of further interest to researchers—in our case those that contained b-quark jets—and "not interesting" events, those that were essentially background noise. In high-energy physics, such a device is called a *trigger*. I knew very little of the details of the physics and I worked on a subproblem of the whole problem. In nuclear collisions energy readings are taken from a calorimeter, which produces a large amount of data. The trick was to build a fast trigger by isolating a small region of the full calorimeter reading for analysis. The end result was an 8×8 region of the sensor array (64 pieces of data) for fast analysis. If the reading was interesting, then the full reading could be taken.

I was given a training file of approximately 150 signal events (these were the "interesting" events, in this case they were "B-jets") and 300 background events (the not interesting events). My job was to build a neural network that could discriminate between the two types of events for 64-dimensional inputs. The way things worked out, I worked largely independently. Bruce and Myron taught me about the data and what little physics I needed to know. My job was to make it happen. I enjoyed this approach. I had a big-picture problem to solve, but the details were left to me. I worked quite hard that summer—I would drive out to Batavia, Illinois, approximately 45 minutes from the University of Chicago, and read my reference books—on neural networks, scientific computing, and Fortran—write code and look at the data. I would meet with Bruce and Myron from time to time to update them on my progress.

In the end, I was successful in building the neural networks software and producing very good results with the training data. The system could correctly identify 95 percent of B-jets while rejecting 95 percent of background events. We then tested the system on an independent data sample and found that it could correctly identify 65 percent of the B-jets and reject 95 percent of background events. Note what this means. If the results were representative, given a reading of a very small region of the sensor array, if we got a positive reading (i.e., a B-jet), then we could be 95 percent sure that what we were looking at was a B-jet. Moreover, we would be able to capture 65 percent of all B-jets that were present. These results were eventually published in a paper I coauthored with Bruce, Myron, and several others.[11]

The University of Chicago PhD Program

In thinking about how I became a quant, I think back to all the little experiences I had and all the work I enjoyed in high school and college and everything pointed toward something that would have looked a lot like quant finance. I enjoyed working with computers. Of all the computer jobs I had, I enjoyed the Fermilab job the most because I enjoyed working with and analyzing data. I had a definite sense that building something that "worked" was incredibly enjoyable. At the time, however, there was no obvious job for me to take after college. I had some friends who went to Wall Street on the banking side, but this seemed uninteresting to me. I didn't want to be a doctor or lawyer so my options—it seemed to me—were to either go to graduate school or try to work at Fermilab.

I enjoyed the work at Fermilab. But I was not a physicist and did not have a PhD and it did not make a whole lot of sense to work there. I decided to get more education. Having narrowed this down, my biggest challenge was whether to do pure or applied mathematics. Looking back here is where I probably made my biggest mistake. I decided to study pure mathematics. At first I pursued an area called *dynamical systems*, which is an applied-seeming area of pure mathematics, I went to Caltech in their first year graduate program, but just as I arrived their dynamical systems expert announced he was leaving. At that point, I applied to University of Chicago and also to Caltech's applied math program.[12] I was accepted at both, but decided to go back to Chicago. I have always wondered what would have happened had I done some area of applied mathematics and why I did not.

For one thing, applied mathematics made a lot of sense. I would have especially enjoyed working on scientific computing and areas where computer implementation played a big role. I wonder if I would have ever left academia. It seemed a lot less otherworldly and it had the sorts of problems I am more interested in and better suited at solving.

Nevertheless, I dove into the deep end of mathematics and decided to not only to pursue pure mathematics, but what I consider to be the purest of pure mathematics. I worked in an area broadly known as the Langlands Program, after Robert Langlands, who is a professor at the Institute for Advanced Study and had made far-reaching conjectures about the connections between number theory and geometry and analysis. My thesis was a generalization of a breakthrough paper by David Kazhdan and George Lusztig of Harvard and MIT.[13] They had proved a small piece of the Langlands conjecture using interesting new methods from geometry. My advisor, Robert Kottwitz, "had a hunch" that it could be generalized to a larger piece of the conjecture. The actual work was appealing to me because it tied together and utilized a broad swath of mathematics, including algebraic geometry, number theory, group representation theory, and topology.

There is one interesting lesson I learned about getting a PhD in mathematics from my experience. From the outside, anyone with a PhD in mathematics would naturally be rated as having extremely good math skills. Broadly speaking this may be true, but there is tremendous variation among PhDs. In the first place, it's not an exaggeration to say that almost no one thinks of their own PhD problem in mathematics. This was especially true in my field. My advisor once went so far as to say "no one in this field has ever thought of their own thesis problem." I was amused by this because one of the most famous results was Tate's thesis (Tate was my advisor's advisor), so I shot back, "What about Tate's thesis?" "Nope, that was Artin's idea [Tate's advisor], Tate just did the work."

So it was my advisor, Robert Kottwitz, and not me who had a hunch that the Kazhdan-Lusztig result could be generalized. This was fairly typical for PhD work in mathematics—the student solves a problem given to him by his professor—but also differentiates the work of a PhD student from that of a world-class mathematician. To be at the truly highest level, you have to either conceive of problems that shed light on important areas of mathematics or solve incredibly difficult problems employ techniques that previously seemed unrelated to the problem.

My time at University of Chicago was incredibly fulfilling and I have nothing but the highest regard for the PhD program and the University in general. Bob Kottwitz was terrific. He recommended a problem that was respectable—in that if I solved it, it would be a publishable result—but not so difficult that I would spend 10 years in graduate school swimming upstream. He also took a very hands-off approach, exactly the approach I like.

Academia

Mathematics has a way of pulling you in. When I actually solved my thesis problem I was extremely excited about my prospects in mathematics. For one thing, I finished my thesis in the winter of 1992. Theoretically, I could have been out of Chicago in two years save for the department's philosophical objections to two-year PhDs. I spent most of the 1992–1993 academic year working with Victor Ginzburg on a new book on the relationship between algebraic geometry and representation theory—a project that would last many, many years but ultimately produce what I consider to be a terrific book on the subject.

In the fall of 1993, I decided to apply for academic jobs. It was the natural thing to do. I had been doing well so far and had a lot of things going. It was a logical next step. I ended up with an offer to go to the University of Toronto for 1993 to1996 and an offer to go to the Institute for Advanced Study in Princeton for the 1994 to 1995 academic year. I was overall pretty excited by my offers because Toronto had a great department for my field and Langlands himself was at the Institute.

So at summer's end in 1993, I packed all my belongings into my car and drove from Chicago to Toronto to start my life as an academic. Toronto is a beautiful city and the university a terrific math department, but it was far from home and I felt lonely and bored. Aside, from my book project with Ginzburg, I was not feeling particularly motivated to do mathematics. I started to think about alternatives. This was the first time I seriously considered quant finance. Here I have to credit my friend John Liew,[14] who I met during college and who I had kept in touch with through graduate school and beyond. John was a natural finance guy. He majored in economics at the University of Chicago and went on to do his PhD in finance at the business school. All through graduate school, John asked aloud why I was not doing finance. So finally one cold, dark, day in the dead of Toronto's winter, I took a copy of John Hull's book *Options, Futures and Other Derivatives* out of the University of Toronto library.

Like many mathematicians and physicists, I found the mathematics of the Black-Scholes options pricing formula incredibly interesting. For starters, after years of specializing in pure mathematics, I was starting from scratch in a totally new area. It allowed me to start to learn basic mathematics instead of delving deeper and deeper into advanced subjects. I literally had to start from scratch and learn probability theory and then the basics of stochastic processes, things I knew nothing at all about. Not to mention I knew nothing about financial markets, derivatives, or anything at all to do with finance. It was exciting to learn so much from scratch. In the midst of reading about Black-Scholes, I was also deeply involved with writing the book with Victor Ginzburg from the University of Chicago. Ginzburg was a first-rate mathematician and a terrific writer. He was also a perfectionist to the highest degree. We would write and rewrite chapters endlessly; and each new theorem we added seemed to inspire the need for new chapters. The book went from a proposed hundred-page set of lecture notes into a long, involved project.[15]

As I was learning about the Black-Scholes formula I was growing increasingly frustrated with the Ginzburg book project. I needed relief. I decided to take everything I was learning about options pricing and write a book on the subject. Now why would I do that? By working with Ginzburg, I had learned to take extremely complicated ideas and explain them clearly and concisely. Also, since I was learning Black-Scholes from scratch, I thought I could bring a fresh perspective to it. I promised myself I would stick only to mathematics and not try to say anything about markets or things I knew nothing about.

I got the idea for the options book some time near the end of my time in Toronto and started the project for real in the winter of 1995 while at IAS. In Toronto, I continued to work on the book with Ginzburg—which we had named *Representation Theory and Complex Geometry*—and work on mathematics research, but I also began to write a paper on options pricing. Thus, my career as a quant slowly began in Toronto in 1994.

I never published that first paper, but I did post it on the Social Sciences Research Network (SSRN.com). It was called "An Options Pricing Formula with Volume as a Variable." The idea was that the Black-Scholes formula relies on perfect dynamic replication of an option with a portfolio of the underlying stock and a riskless security. My idea was to ask, what if instead of perfect replication you can only replicate with a certain probability? What I did was show that if you could replicate a security with another with an arbitrarily high degree of probability, then you could obtain pricing formulas that had all the good properties associated with perfect replication. In the paper in question, I showed that you could price options assuming the underlying stock followed a Poisson process. Later I showed with Michael Ong that this could be used to "correctly" price binary options using call spreads.[16]

I wanted to get feedback on my paper and so I sent it around to various academics and also to Emanuel Derman, who was head of Goldman Sachs' Quantitative Strategies Group, which was the leading group in derivatives pricing research at the time. Derman's group was in charge of derivatives research and technology, and he was the most recognized figure in the industry at the time. One day, while sitting around IAS, Emanuel called me to tell me he had received my paper. He told me that he had a regular seminar invited me to come in and give a presentation to his group on my paper in an informal seminar he ran. I was extremely flattered and gladly accepted.

Giving the talk at Goldman was nerve-wracking for me. I was more nervous than I had ever been giving any lecture before. This shows how much context matters in life. I had lectured in front of the entire IAS mathematics faculty and had given a talk at Harvard's mathematics seminar in front of various Fields Medalists and had never been nervous. In those talks I felt I knew what I was talking about and was on solid ground. Here, I was giving a paper in an area I did not know well in front of a group of experts.

Fortunately, the group was friendly and quickly put me at ease. My paper was of no practical use. Nevertheless it did have some (in my opinion) interesting mathematics in it. After the talk, Iraj Kani, Derman's key research collaborator, and some of the other people from the group took me to dinner. At the dinner, Iraj proposed that I interview for a summer job in the group if I had any interest in leaving academia one day. This was perfect for me because it gave me an opportunity to try my hand in finance without having to leave mathematics altogether. In addition, as I was writing my book on Black-Scholes, I was beginning to realize how little I knew about either its practical application in a business context or how it was actually implemented. Further, the timing for the job happened to work out perfectly.

While I was at IAS, I decided I did not want to return to Toronto and I applied for jobs yet again. This time I got what I wanted. I received a National Science Foundation grant and an offer to come to Harvard University in the fall of 1995. Given this, I had the perfect situation. I was finishing at IAS in the spring and had the summer off.

I interviewed for the job and I was fortunate that they chose me—I think mainly on the strength of the work I had done at Fermilab and my interest in actually implementing financial models. Derman and Kani had a particular task in mind. They wanted to extend some of their key option pricing work from binomial trees to trinomial trees. In 1994, Derman and Kani had caused a stir in the options world by publishing a paper that showed how to fit a binomial tree to price all options currently trading in the market correctly.[17] My job was to extend their work from binomial trees to trinomial trees and implement it into their option-pricing library (which was written in C code). The theory part was an easy extension of what Derman and Kani had already done. The real work was writing the code. This was fine by me, because I wanted to see how a real options-pricing system worked, and I could think of no better place than Goldman Sachs to do it. In the end, the job worked out quite well. I implemented the trinomial tree model in their system and we published a paper on it.[18]

The summer was a busy time and I barely got to enjoy New York. I spent my days working at Goldman and nights writing my Black-Scholes book and *Representation Theory and Complex Geometry.* In addition, I worked out a new bit of options theory that extended Derman and Kani's binomial tree work to American Options,[19] which was probably the only example of actually useful options-pricing work I worked out on my own. An additional benefit of working at Goldman for the summer was that with all the new theory I learned at Goldman, I felt qualified to expand my book to include the newer material on options pricing developed mainly by Derman and Kani, and hence decided on the name *Black-Scholes and Beyond* for my book.

The Harvard Mathematics Department

Harvard was the beginning of the end for me in mathematics. I began my postdoc at Harvard in fall 1996 after spending the summer at Goldman Sachs. On the one hand, Harvard was a dream come true. I was thrilled to have a position in Harvard's mathematics department. To my mind it was the most exciting department in the world. If I was going to be doing mathematics, I wanted to be at Harvard. In practical terms, however, I had a few issues to deal with.

First of all, I was in the final stages of finishing *Black-Scholes and Beyond*. I was pretty sure I would have a published book by late 1996.[20]

I was still working on *Representation Theory and Complex Geometry*, and there was some hope of finishing it by the end of the 1995–1996 academic year. I was also involved in a very interesting research project with Joseph Bernstein, a former Harvard professor who was now working at the Tel Aviv University. I was getting spread too thin. In particular, I realized that I was in the honeymoon period for academia. In mathematics, postdoctoral positions are a great deal. You have a light teaching load and are left alone to do research. Very quickly, however, you have to get a tenure track job which means joining a mathematics department with all its committees, a bigger teaching load and more real responsibility. The way I saw it, I had two more years on my grant at Harvard and, if I wanted to be a real mathematician, I would have to commit now to mathematics. On the other hand, I could leave Harvard and work on Wall Street on quant research. As I saw it, those were my two major options.

I decided to explore leaving mathematics. I put together my resume and began sending it to people I had met on Wall Street. It was actually an unexpected offer of a promotion to assistant professor that finally settled things for me. At that point I was a postdoctoral fellow—a fine position, but an assistant professor was in every way better. It was no big deal from the faculty's point of view. It really meant that I would teach calculus. But to me it was a form of validation. On the practical side, it meant I could stay three more years in a higher position at my dream mathematics department. It was ironic that the very fact that this was so persuasive to me—that external validation meant so much to me—was actually proof that mathematics was not right for me. If I thought about Hilbert or my archetypal ideal mathematician, the only difference this should have made was that it would allow me to stay in a great place to do mathematics for a longer period of time. Put more simply, I should have loved the opportunity, not the prestige. Once I realized this, I knew it was time to move on.

Moving to Wall Street

My move to Wall Street was a gigantic shift in my entire way of life as much as it was a geographic one. I moved from Cambridge to Manhattan, and I moved from a math department to an investment bank. The move represented a shift away from mathematical research—freedom to explore whatever problems I liked—to quant research within a large investment bank where I would work for someone else doing what they wanted me to do. I had some trepidation about this, but I was confident in my decision.

My exit from Harvard could not have gone better. When I had told my scientific advisor Benedict (Dick) Gross that I was leaving Harvard to go to Wall Street I had been extremely nervous. I literally had no idea what to expect—disapproval? banishment? My fears were naturally overwrought. Dick invited me to lunch at the faculty club and talked to me about his view of being a mathematician. This might have been the most important mathematical conversation of my career.

Dick had obviously thought about this before. He told me that the notion of a full-time mathematician was actually a fairly a modern one and that historically mathematicians had more interaction with the real world than today. We discussed Karl Friedrich Gauss—one of the all time great mathematicians—who was commissioned by King George IV of England to survey the Kingdom of Hannover in 1817. Gauss was a great pure mathematician and believed in the preeminence of pure mathematics. He once said, "All the measurements in the world are not worth one theorem by which the science of eternal truth is genuinely advanced." Yet Gauss's work in surveying led to great discoveries in mathematics, data analysis, and the science of surveying.

While surveying Hanover, Gauss developed new theoretical methods of fitting data containing errors using ordinary least squares and invented the heliotrope, a surveyor's instrument that represented a vast improvement over tools available at the time. Gauss's work as a surveyor also inspired him to think in new ways about the foundations of geometry which led him to develop the field of differential geometry, a field of central importance to this day.

Dick and I talked for a long time and his message was similar to the conclusion I had come to. There are lots of interesting things to do out there and mathematics is really difficult even for great mathematicians. So, if you don't love it, there are better or at least equally valid things to do with your life.

With Dick's blessing, I officially took a year's leave of absence with the understanding that I could return if my year away inspired me to be a mathematician. His advice was both an inspiration and a model for how to be a great mentor. Dick was honest, reflective, and had my best interests at heart. I am very thankful I had the opportunity to speak with him before leaving academia.

Quant Research

The actual job I got was in the quant research group in Morgan Stanley's Institutional Equities Division. I was hired to work on models of portfolio trading for their cash equities program trading desk. My boss explained to me that the program trading business was looking for ways to utilize quantitative techniques to improve their business. I was extremely excited about this! I liked the idea of working on something completely away from derivatives. In truth, I was not sure I wanted to go down the road of doing derivatives. I felt it was a well picked-over field and I wanted to work on something new.

When I joined Morgan Stanley, I tried to learn as much as possible about the business of program trading by spending time with the guys on the desk and attempting to understand the business in order to model it. The basic idea was simple enough. The program trading desk would receive orders to trade portfolios of stock, portfolios that were too big to trade in a single order and that had to be worked over the course of the day. The desk would carry out these orders while managing the risk of the positions and trying not moving markets too much with their trades. They profited in one of two ways—either from commissions or from doing the trade as a principal and trading out of the position for a total cost of less than what they paid for the positions.

I found the business incredibly interesting and was extremely interested in building a model that would help make the business run more efficiently. But, shortly after I joined, my boss, the head of program trading, and his boss resigned en masse to join another firm. On the bright side, I had learned enough and had produced a first model for the program trading. I was disappointed at the turn of events. I had enjoyed working with the program trading team. But the silver lining for me was that I ended up publishing the work and the model itself has become so widely used as a basis for algorithmic trading.[21] At the time, I had hoped to develop a model and implement it at Morgan Stanley. That is not how it worked out, but in the end it was a great introduction to quant research.

For a first job on Wall Street, things went pretty well. I learned a great deal and I worked at a truly first rate firm. I also met a number of people who have had a tremendous influence on my career. One person in particular was Peter Muller. Peter was running a proprietary trading group that was considered by everyone to be the crème de la crème of the equities division. I did not know—and still do not know—exactly what they did, but whenever his group was mentioned, people were abuzz with excitement. Peter was outgoing and friendly and made an effort to get to know the quant research guys. We became friends and he was eventually a big influence on my decision to pursue quantitative trading.

Quant Research and the Mathematics of Portfolio Trading

I want to start with some general comments about how I view research in financial mathematics. I believe that one incredibly important application of quantitative research is in improving business processes. Many business units in finance are governed by rules of thumb and best practices that have never been fully analyzed. I believe this was certainly the case in program trading. There are undoubtedly many more problems that are not yet seen as problems in this light. At the time I joined Morgan Stanley, it was not really a research problem to ask whether there is a mathematically optimal way to trade portfolios. I cannot say whether other banks were asking this question, but I can say that Morgan Stanley was and I was fortunate to be asked to work on the problem. I think the search for interesting problems in financial mathematics should begin with a look at what businesses are out there and how they operate. One should ask whether this business can be analyzed from a mathematical standpoint and thereby improve the business.

Once one identifies the problem, the next step is clearly to solve it. Here I believe the best approach is to search for the most appropriate mathematical setting in which to cast the problem. This sounds obvious, but in fact much research in financial mathematics proceeds in the opposite direction. Researchers start with a certain set of tools and search for problems which can be solved with them.

In the case of Optimal Liquidation, the basic idea is that the trading of a portfolio of stocks traces out a path in time according to the amount of the portfolio which remains to be liquidated. The notion of looking at a particular way of trading a portfolio as a path, or trajectory, was the key insight because once you look at it that way the natural question to ask is, which trajectory is the best one? At this point the problem is reduced to two subproblems. First, how do you decide what makes one trajectory better than another, and second, how do you find the best trajectory? Fortunately, there is mathematics ready-made for this setup.

At each point in time, the stock portfolio has two costs associated with it: a risk cost and a market impact cost. The *risk cost* is the theoretical cost associated with holding a risky position that you do not want to be holding. The *transaction cost* is the cost associated with the market impact of the position's changing through time. The total cost is the sum of the transaction costs and the risk cost appropriately adjusted by a risk aversion parameter, which controls for how urgently you want to reduce the risk.

When you look at the problem this way, it naturally fits into the mathematical framework of the calculus of variations. This is a well-understood optimization problem whose solution is given by the *Euler-Lagrange equation.* This was the basic idea—to model program trading as an optimization problem over trading trajectories.[22] The details lay in how to measure risk and cost. The important point is that once you understand the problem fully, the most appropriate mathematical solution is obvious. It's important to distinguish this from the opposite approach, which is to start with a known set of tools and try to find problems that can be solved by them.

This was the original model I worked out at Morgan Stanley. Later, I started collaborating with Robert Almgren, who was then at the University of Chicago. We met when I gave a seminar on the model at U. of C. and we began talking and it became immediately clear that we had tremendous research chemistry. Together, we extended the model in a variety of significant ways, analyzed it more detail and in late 1997 circulated our original paper on the subject.[23] In that paper, we extended the basic idea to study the entire range of possible trajectories that were optimal for different levels of risk aversion and studied them together, thus establishing what is now referred to as the efficient trading frontier. Since then, Rob and I have continued collaborating and he has continued to do important research on optimal trading. He is a truly first-rate mind and has been a fantastic partner in mathematical finance.

Quantitative Portfolio Management

There came a point in time when I realized that if I was going to stay in finance I had to move to a revenue-generating role. This is not a must-do for quants, but it was a must-do for me. My issue was that one of my personal drivers was the need for my work to relate directly to the main purpose of the business. Looking back at academia, I liked the fact that while I was a research mathematician I worked in mathematics departments and the main purpose of the mathematics department was to produce research (yes, and of course to teach). I was doing the most important work as far as the mathematics department was concerned. When I went to Wall Street, I thought, I do research in academia, so I will do research on Wall Street. This is one way of making the analogy. But a better way would have been to say, In academia, I am doing the most important work of my department. I should do the work that is most important to whatever organization I work for.

The fact is, Wall Street is about making money. It's hard to see it any other way. Certainly the great Wall Street firms care about their clients and want to provide first-rate service and products, but it all boils down to producing revenue. In this respect quantitative research, as I was doing it at Morgan Stanley, was at least once removed from the core activities of the firm. After much thinking, I decided to move into portfolio management. I was fortunate to get a job with Goldman Sachs Asset Management (GSAM) in its Quantitative Strategies group.[24] I was hired as a portfolio manager and research with the mandate to work on the overall portfolio management activities of the group and to develop a new trading strategy.

The experience was eye opening to me. First of all, it combined pretty much every aspect of what I had enjoyed doing since I was double-digits in age. I was doing research, but it was directed at making money. I had a direct hand in implementing these research ideas into a system which required programming. The research involved a great deal of data analysis. In short, it was everything I had always enjoyed.

A particularly appealing aspect of money management was the ample need for technology. I have always thought that having a strong background in technology—having had the experience of writing tens of thousands of lines of code—was an advantage. To do anything large-scale in quantitative finance requires good technology. Everything from the storage and retrieval of trade data to the execution of trades increasingly relies on technology that is efficient and reliable. Knowing how to build systems that work well now and can be maintained and grown is an important asset because as a business grows in complexity, the technology that drives it has to grow with it. It is all too easy to build a small system that works well but find that as it grows it grows out of control in complexity and becomes difficult to manage.

Mathematical Finance Education

On a final note, I want to talk about my involvement in the development of financial mathematics education. During the summer of 1995 when I worked at Goldman Sachs, Niels Nygaard from the University of Chicago asked me to help research the curriculum for a new master's degree program in mathematical finance that U. of C. was planning. The project had been proposed in 1994 by Bob Zimmer, the department chairman at the time, who had noticed that many PhDs were moving to Wall Street. In a prescient move, he motivated the start of the program and through my relationship I got involved.

From the start, without any inkling of what the next 10 years would bring, I thought this was a great idea, and I thought Chicago would be a terrific department to develop and run such a program. As it turned out, when I moved to New York on a permanent basis, I was contacted by New York University's Courant Institute of Mathematical Sciences, where I learned that it also starting a program. I was asked to be the program's director on a part-time basis while continuing my day job as a quant.

The Courant program grew out of two mathematical finance courses that were taught in the early 1990s by Marco Avelleneda at Courant, and the idea was to build a program around these courses. The push to do it came from Courant's then-director David McLoughlin—now NYU's provost—who I think, like Zimmer, saw the wisdom in supplying training in core subjects in the mathematics of finance. My job would be to help form the curriculum, administer the program and recruit adjunct faculty to teach practitioner courses. I was flattered to be offered the opportunity and to work with Courant's faculty, including Robert Kohn and Jonathan Goodman, who were instrumental in getting the program off the ground. I also worked closely with the program's chairman of the board, Jeff Rosenbluth.

Reaching out to expert practitioners on Wall Street appealed to me as well. There has always been a part of me that liked to recruit people into doing things. Even as an academic I would try to sell people on new research projects and get people involved. During my first year at Courant, I was very fortunate to recruit three world-class people to teach. Jim Gatheral—a managing director at Merrill Lynch in the equity division in quantitative research,[25] agreed to teach a basic markets course with me. We had a fantastic time doing this and have remained friends to this day. Steve Allen, who was then head of derivatives risk management at Chase, agreed to design and teach a risk management course, and Peter Fraenkel, who ran a derivatives pricing and hedging technology team at Morgan Stanley, agreed to teach a course called Computing in Finance. Together, Jim, Peter, and Steve formed the core of the people who taught the practitioner courses.

My second year I recruited Nassim Taleb—of *Fooled by Randomness* fame—to join us. Nassim and I had met through a random exchange of e-mails after I completed *Black-Scholes and Beyond,* and I admired his bold style and contrarian thinking. I invited him to coteach a course with Jim Gatheral on "anything they would find interesting to teach." Together Jim and Nassim devised a course called Case Studies in Finance, which has become a staple of the program and enormously popular. They have taught it every year since then.[26]

I worked at Courant from 1997 through December 2002, when I resigned my position mainly because of the increasing demands on my own life. When I started, I devoted most of my free time to running the program. At the end of 2002 I took a new job as managing director of Quantitative Strategies at SAC Capital Management. I felt I could no longer devote the time I previously had to the program and decided that I could not devote the amount of time and energy to the program that I had in the past. This also gave me an opportunity to reconnect with University of Chicago. In 2003, I agreed to work with the University of Chicago Financial mathematics program as its executive director.[27]

One of the most exceptional things during the decade of my involvement has been that the financial mathematics degree has become an accepted professional degree. From the very beginning, we all believed that for financial mathematics programs to succeed, we would need to see the emergence of many such programs. After all, would a Harvard MBA be a Harvard MBA without 1,000 other schools offering the same degree? I think, as a result of the success of the early programs and the demand for their students, there was a slow, steady decision in many mathematics departments to develop their own financial mathematics programs. Today, the degrees are well known to recruiters, human resource personnel, and most of the top investment banks and hedge funds.[28] As of this writing, there are hundreds of such programs worldwide.[29]

Final Thoughts

I officially left academia in the spring of 1996. It's been close to 11 years since I entered the world of finance. In thinking about those years, there is no doubt these have been exciting times for quants. The world of finance has become more technological and more quantitative. This has happened for a number of reasons, having both to do with financial markets and instruments and with technology developments. On the markets and instruments side, the "electronification" of many financial markets, the development of important new financial instruments and markets including credit derivatives and energy derivatives, and the increased attention to risk management have created more demand for quantitatively trained people. As such, we have also seen leakage of quantitative methods and thinking to traditionally nonquantitative types of trading and money management. This has had much to do with the increased availability of financial data, cheap computing power, and increased flow of information through the Internet. In all, the level of quantitative sophistication and the utilization of technology in finance have risen dramatically during the past decade.

On a personal level, working in financial markets and, in particular, money management has been a fantastic choice for me. I never regretted my time in academia and feel fortunate for the experiences I had. The time I have spent in finance has also coincided with the great increase in the number of hedge funds and assets managed by hedge funds. In particular, some of the largest, most successful hedge funds have some or all of their assets managed using quantitative methods and a heavy dose of technology. This does not entirely surprise me and is probably the biggest single change I have seen for quants since I left academia in 1996. At that time hedge funds were barely visible to mathematicians. Several large hedge funds were recruiting mathematicians and physicists, but nobody I knew really understood anything about them. They seemed outside the norm. Today they are the norm.

For quants themselves, one of the biggest changes I have seen is one of perspective. Math and physics majors looking to move into finance today have a completely different outlook and range of options than they did a decade ago. In the early 1990s, most of the people I knew who went into finance started off going to graduate school making an earnest attempt to become academics. The ones who move to finance did so either because they discovered finance and found it an exciting alternative or because they discovered academia was not for them. Today, things are quite different. First, because of the rise in popularity of mathematical finance programs such as University of Chicago's, many would-be math and physics PhDs are getting masters degrees in financial mathematics and going directly to finance. Second, many PhDs and academics looking to leave academia are going directly into quantitative portfolio management and not quantitative research. This is not a small change. It's a change in perspective from I want to do research for the firm to I want to trade and make money for the firm.

I think that this change in perspective, which is something I have seen broadly in the last 10 years, is extremely good for financial markets. It may have two long-term effects. First, it will help bring effective use of technology and risk management into all aspects of portfolio management. It is increasingly obvious that quantitatively based asset management has been extremely successful and that at least some of this is due to effective use of technology and risk management. As quantitative and technical people become an increasingly important part of the landscape, I believe we will see portfolio management teams that are increasingly populated with a blend of quantitative and nonquantitative teams. If a fundamental long-short portfolio manager can more effectively manage his or her positions and risk with better technology then it is inevitable that this will be adopted. Second, it will bring an increasing number of extremely quantitative people into traditional areas of portfolio management. Their approach will most likely be more analytical and closer to "quantitative" than traditional managers, thus blurring the lines between quant and nonquant.

Finally, I think that quantitative asset management's range is going to increase dramatically over the next 10 years. Today when people think about quantitative asset management they usually think about statistical arbitrage or global tactical asset allocation. Over time, I see quantitative methods being applied to an increasing range of products. Driving this will be increased liquidity in new markets, the availability of data to analyze and the availability of electronic access to those markets.

Chapter 8

Peter Carr

Head of Quantitative Financial Research, Bloomberg

I'm thrilled to be asked to describe how I became a quant. As the mythical Ted Baxter intoned, "It all started in a 5,000-watt radio station in Fresno, California. A $65 dollar paycheck and a crazy dream." For me, it all started the day I took my first job as a paperboy, delivering the *Toronto Star* in the early 1970s. I realized that you could deliver a lot more papers to people in apartments than houses, and so I cornered the paper delivery market on every apartment building between my home and school. I figured that if I have to walk home from school everyday, I might as well earn some cash.

At first, that cash went into my savings account at the local branch of the Canadian Imperial Bank of Commerce. Then I realized that I would have better access to my meager savings if I split it among strategically chosen branches of the big five Canadian banks. That meant I didn't need to carry around as much cash in my wallet and I could maximize the interest earned in the accounts. In those days, there were no ATMs and no minimum balances, and so I probably had a dozen bank accounts with a dozen dollars each.

I thought it was a miracle that banks would pay you interest on their interest, and so I saved every penny I could. I noticed that the Canadian banks let you have accounts denominated in U.S. dollars (USD), and that the American interest rate was higher. So I opened yet more accounts and kept my eye on the exchange rate (printed every day in those papers I was delivering). In those days, the Canadian dollar (CAD) was about par with the American dollar, but the Canadian dollar was weakening. The papers published the exchange rate both ways—that is, CAD per USD and USD per CAD. As the Canadian dollar fell from, say, one U.S. dollar to 99 cents, the papers showed USD per CAD rising from 1 to 1.01. My father thought that if the Canadian dollar dropped further to 90 cents, then USD per CAD would rise to 1.10. I asked him what would happen if the Canadian dollar dropped to zero, and I was promptly grounded for my curiosity. My father is an electrician who pulled himself up from poverty in Malta to a comfortable existence in Southern California today. He's a very smart guy, but he was pulled out of school in grade 6 to work as a conductor on his family's bus. Sitting alone in my room, I thumbed through the Grolier encyclopedias, and happened upon the Pythagorean theorem.

My First Eureka Moment

In my first eureka moment, I realized that my father could use the Pythagorean theorem to figure out where to cut wire when running it diagonally. When I explained it to him, he was so happy that he let me out of my room. I was very impressed that math could be used to solve problems (in this case, the problem being that I was stuck in my room).

I spent high school trying to figure out how to get a girlfriend, never very successfully. Although I was team captain of the school's entry in a trivia contest, called "Reach for the Top," that didn't seem to impress any of the girls I was interested in. About the only way I could get a girl to talk to me was to get both of us drunk. The drinking age in Canada at the time was 18, and that meant that alcohol was freely available. I stopped drinking when I turned 18, and I don't drink at all today.

I studied accounting and economics as an undergraduate at the University of Toronto (U. of T.). Unfortunately, they are about the worst subjects you can learn if you want to become a quant. Luckily, the requirements for my bachelor of commerce degree required a single finance class taken in your third year. That finance class was taught by a gifted professor named Dan Thornton who emphasized the use of physical analogies to explain the math. For example, he used Modigliani Miller's explanation of the irrelevance of the capital structure by showing that a farmer can't expect to make money in equilibrium by separating milk into cream and skim.

I did well in that finance class; but I got sucked into the auditing vortex when I graduated. As in America, every public company has to have its financial statements audited, and in those days prior to the personal computer, auditing was a very labor-intensive activity. The big-eight accounting firms (now the big four) more or less minted money by hiring unsuspecting undergrads and charging them out at a huge multiple of what they were paid.

Accounting for the Future Instead of the Past

After ticking and bopping for a year, I got tired of the abuse and decided to get an MBA from U. of T. With my undergrad degree in commerce from the same institution, I was able to get a two-year degree in just a year. In that year, I had the good fortune to be taught finance by another gifted professor named Laurence Booth. He would teach the exact same class twice, and I sat in on both sessions.

I loved the theory of finance and decided to get a PhD in it, simply because I thought it was fascinating. I had no idea what business school professors earned, which is good because at the time it was about the same as other academics. I consulted with Dan Thornton and Laurence Booth, and they wrote my letters of recommendation. Dan Thornton let me in on an important secret, which is that academics use English words to describe mathematical concepts and that one should never confuse the everyday meaning with its mathematical definition.

Tired of the dreary Canadian winters, I applied only to Californian schools and decided on UCLA because it had some established professors such as Richard Roll. Although classes started in September, I arrived at UCLA in May and caught a finance seminar on Ross's APT (advanced portfolio technology). I was completely blown away with the level of mathematic rigor in that workshop, and so I spent the summer sitting in on undergraduate math classes, desperately trying to make up for lost time. At that time, I bought my first math book that I didn't need for a course. I now own thousands of such books, and they are my crutch for a lack of formal training.

In retrospect, I was lucky with my choice of UCLA on several counts. First, it was primarily an empirical school, nicknamed University of Chicago at Los Angeles by the cognoscenti. That meant I could just keep up with the math, which I'm sure I couldn't do had I gone to Berkeley or Stanford. Second, UCLA had some great assistant professors such as Grinblatt and Titman, who cared about doctoral students and were passionate about what they did. Third, UCLA later hired Brennan and Schwartz just as I got into the research phase. I particularly admire Michael Brennan, a gentleman and a scholar, with the broadest vision of finance I have ever seen. I sat in on an MBA class he taught and was enthralled by his wisdom. Unfortunately, his impact on other students was not so great because the only feedback he got from his UCLA MBA students was that he should consider wearing a different tie to class.

Postdoctoral Studies

Wall Street was just beginning to hire quants when I graduated from UCLA in 1988. I hate wearing suits, and so academia was the only choice for me. I agonized over which university to go to and chose Cornell because I liked the idea of being in an isolated place where I could continue to catch up.

While I was at Cornell, I consulted for several firms such as Susquehanna Investment Group on exotics, Astro Gamma, which made FENICS, and Mobil on real options. I really liked the opportunity to turn theory into practice, and it was great to have Bob Jarrow and David Heath around to run ideas by. In my last year at Cornell, Dilip Madan spent his sabbatical year up there, and that changed my life. Dilip is the most amazing guy I have ever met. He's a human light bulb, a device for turning energy into light.

When I started working at Morgan Stanley (MS), I brought him in as a consultant and we hit on several ideas in rapid succession. I could see that variance swaps were primed to take off, and I left MS for Bank of America to help make it happen. When Bloomberg decided to start up a high-powered quant group, I joined them to help make it happen. I really enjoy interacting with my talented colleagues, some of the best quants in the business. Our vision remains one of putting sophisticated analytics in the hands of 260,000 users.

And in the End ...

To wrap up, I'd say that I've had my share of ups and downs over the years, as have many quants I've known. The one constant that keeps me in the game is my love for the subject.

Thanks to the blindingly brilliant insights of intellectual heavy-weights like Robert C. Merton and Stephen A. Ross, we have a powerful framework with which to tackle fascinating problems. Although many academics think otherwise, mathematical finance is in its infancy with lots of low-hanging fruit to savor. If you want to be a quant, my advice is to see it as a calling, not a job. The money is incidental. It comes and goes. A good idea never goes away and if you are lucky, you'll have a few before we all fade away into obscurity.

Chapter 9

Mark Anson

CEO, Hermes Pensions Management Ltd. CEO, British Telecommunications Pension Scheme

Recently, I gave a PowerPoint presentation to an audience on private equity returns and benchmarks. As part of my presentation, I included several slides demonstrating multivariate analysis of several potential benchmarks for private equity. As I put my first slide up on the screen, I prefaced my comments by saying, "And here, I begin with a simple equation." My comment brought laughs from the audience for they did not believe that my equation was all that simple. It is moments like this when I realize that I am a quant.

Whether we are called quants, quantjocks, gearheads, computer monkeys, or other affectionate/pejorative terms, the fact is that quants think differently than the rest of the world. We are forever trying to explain economic phenomenon through the use of empirical techniques. This separates and distinguishes us from the rest of the investing profession. Although I might have thought that my multivariate analysis was simple—and indeed, to a quant it was simple—the rest of the investment audience considered it complex.

Yet, it need not be this way. I am less concerned with dazzling audiences with my quantitative skills or technical jargon and more concerned with understanding the empirical relationship between economic variables. This is why I became a quant. I want to be able to ask and answer the questions *why* and *how*. Why, for example, do two economic variables such as late-stage venture capital returns and small-cap stocks move together in a consistent fashion? How strong is this relationship? Can it be used as a reliable benchmark by which to measure manager skill? It was a deep-seated desire to understand the nature of financial phenomenon that led me to become a quant.

PhD, Why Not?

Still, I became a quant somewhat by serendipity. Fate takes a hand in every person's career, and fate played an early role in mine. As I applied to graduate business schools more than 20 years ago (was it really that long ago?—It seems like yesterday), I used the traditional method of applying to MBA programs and then visiting the universities where I had been accepted. It was on such a visit to Columbia University in New York City that the assistant dean for the MBA program—later the dean of the PhD program—suggested that I might make a good PhD candidate.

I was surprised by the suggestion, and more than a little flattered. The idea intrigued me. I truly wanted to go beyond what I call *textbook learning* to delve deeply into the problems of the financial markets, where standard MBA textbooks did not venture. Furthermore, I had written an undergraduate economic thesis where I prepared a case study of inventory models on a local manufacturing firm near my college. Consequently, the idea of independent research appealed to me and was already consistent with my background. Perhaps that is why the dean thought I might make a good PhD candidate.

From my point of view, I had nothing to lose. I had already been accepted to the MBA program, so I thought: "Why not?" I took the

PhD application home with me after my visit, spent the better part of two weeks filling out the extensive essays, and submitted my application to Columbia Business School's PhD program in finance. Mostly, I did this as a one-off trade, not expecting to be accepted. I'm not certain who was more surprised—Columbia University when it decided to accept me, or me receiving my acceptance letter. And so, I became a quant, but it wasn't easy.

As I look back now, I realize that I was a bit naïve as to the level of quantitative ability that it took to acquire a PhD As an undergraduate, I had majored in chemistry and economics, so I thought that I had a good grounding in quantitative and economic skills. But this was just enough to get me through my first year in the PhD program. Had I known better, I would have first acquired a master's in mathematics—this would have been the best training to precede a PhD in finance. Still, I made it through that first year and my learning curve improved dramatically.

The key lesson that I learned is that mathematics is pervasive in all forms of economic analysis. When I tell people that I have a PhD in finance, they tend to think of the trading dynamics of the stock and bond market without realizing that a very large quantitative tool kit is required to achieve a PhD Look at any trading desk in a hedge fund today or an investment bank and you will find PhDs by the bucketfull with their doctorates in traditional fields such as finance and economics, but also, mathematics, physics, and engineering. The fact is that many economic relationships follow consistent patterns that can be described through quantitative analysis. When viewed in this light, mathematics, and the other quantitative sciences are nothing more than a tool to describe what we observe in the financial markets.

These quantitative tools can be as blunt or refined as we wish to make them. Consider my example of finding appropriate benchmarks for private equity returns. If a simple one-period linear regression model is used, the alpha, or intercept term, for venture capital with respect to a regression on a public securities index, will be large. However, if one includes other variables in the equation, the alpha erodes as a more refined economic relationship between venture capital returns and the public securities markets is revealed.

Further, one can look at venture capital investments as call options on the future success of the new ventures. The venture capitalist makes an investment in a startup company. If the company fails, the venture capitalist loses her initial investment. However, if the company is successful, the venture capitalist enjoys all of the upside. This simple explanation of the rewards to venture capitalists also describes the payout function for a call option—the loss of the option premium in return for enjoying all of the upside. Recognizing this relationship, and armed with a few option-pricing tools, a quant can fashion a new, nonlinear benchmark for venture capital performance.

Legal Arbitrage

Although I became a quant through Columbia's PhD program, I did not put my quantitative skills to use immediately. At least, not in the traditional sense. Instead, after collecting my doctorate, I went on to law school (and forever have borne the jokes from friends and family about being a professional student). However, even there, my quantitative background served me well because I had learned to approach problems and issues with a keen analytical eye. This allowed me to sort through legal briefs, case law, statutory regulations, and the like in a most efficient manner.

For example, in empirical research, we often talk about *noise in the data.* A good researcher learns to spot the noise quickly and knows how to apply robust statistical techniques to filter out the true signal to test its economic significance. Well, believe me, in the law, there is an awful lot of noise in the data. Sometimes, reading a court case, it seems like there is nothing but noise. But the analytical training I received from my PhD course of study provided me with the tools to filter out the noise and focus on the true signal that judge in a court case wanted to deliver.

As another example, it was while I was a young attorney practicing law in a white-shoe law firm that I figured out how to arbitrage the cost of different legal search engines. There were two computer search engines—Lexis and Westlaw—that contained every state and federal court case then available online. It was fascinating to learn that the text of most current court decisions in America had been recorded in computer files. Lexis and Westlaw both offered computerized access to these court decisions—much easier than the tedious procedure of looking up court cases in bound volumes. While each service had its own special language to access its respective search engines, the services provided were essentially the same—a ripe opportunity for arbitrage.

I quickly learned that each service offered free time for demos and tutorials and that the results of one research project from a free demo with one search engine could be saved in a word-processing document and resubmitted as a new search within a new tutorial with the competing service. By going back and forth between the services, I was able to carry out extensive legal research at a pittance, much to the appreciation of the law partners. Now, you may not need a PhD in finance to figure out this arbitrage, but my mind had already been trained to look for such inefficiencies in the market—regardless of whether that market was financial, legal, or something else.

However, I discovered that there was a trade-off between practicing law and being a quant. Quantitative skills are like a foreign language. The more you use the skills, the more honed they become. Conversely, failure to apply the skills on a regular basis leads to a slow dissipation. Even though I had developed a strong quantitative tool kit, my skills began to erode as my legal career progressed. In the end, I made a difficult decision and left the practice of law to return to the financial markets. I still use both skill sets, but I have found my quantitative background to be the more valuable in the financial markets. There are still a lot of lawyers out there and not as many quants.

Furthermore, when I reviewed the breakthroughs in the financial markets, the more recent breakthroughs are quantitative rather than regulatory. For example, the regulation of the U.S. financial markets was shaped over 70 years ago by the Securities Act of 1933 and the Securities Exchange Act of 1934. These were followed by the Investment Company Act of 1940 and the Investment Advisers Act of 1940—more than 60 years ago.

Conversely, consider all of the economic models that have been developed in the last 40 years: the Capital Asset Pricing Model of Sharpe (1964), Lintner (1965) and Mossin (1966) shaped how assets are valued. The Black-Scholes Option pricing model of 1973 and Merton's Rational Theory of Option Pricing in the same year forever shaped how derivative markets are priced. And there are many other models that have been developed in the last 30 years: Ross's Arbitrage pricing model of 1976, the Fama and French factor models of the 1990s, Grinold and Kahn's Fundamental Law of Active Management in 1989 and its subsequent refinement in 2002 (for transfer coefficients), and the Black-Litterman model of 1990 for asset allocation. The economics of the financial markets are still being discovered and defined, and it is the quants who are leading the way.

Managing the Outcome

As I look forward to new economic challenges in the financial markets, the single largest issue globally is the growing pension crisis. In the United States, the Social Security system faces bankruptcy sometime around 2050. Many defined benefit plans are shutting down or being handed over to the Pension Benefit Guaranty Corporation. In addition, the aging Baby Boomers are finally getting ready to retire and head off into a well-deserved sunset. Further, around the world you see an aging population. In Italy, the median age is 53; in Japan it is 54. The nexus of all of these demographic shifts will place a keen focus on asset/liability management as pension benefits/liabilities grow.

This is on top of the fact that the longevity of an average worker today is much greater than it was just 20 years ago. Advances in healthcare, nutrition, and education have extended the lifecycles of hundreds of millions of people around the world. I see this as the next great area for quants to apply their skill, expertise, and experience.

New investment strategies will have to be invented to deliver stable income and inflation protection for this growing base of retirees. Liability Driven Investment (LDI) is the new buzzword around pension funds. Immunization of a growing liability stream is a key concern for all pension fund managers. This will require more creative solutions, particularly via the derivatives market. Inflation swaps will have a growing importance in managing pension fund liabilities, and quantitative tools will be needed to construct and price these instruments. In addition, longevity risk for pension funds remains unhedged. This is a ripe area for financial engineering.

Further, I see changes for the retail markets, where I expect that there will be new product offerings that are very different from those that investors see in the mutual fund world today. Style definitions such as large cap versus small cap, growth vs. value, emerging vs. developed, and so on, will play less of a role as Baby Boomers retire. In the past, mutual fund products were designed primarily for wealth and savings accumulation.

However, looking forward, I believe that investment products will be focussed less on style management and more on achieving specific outcomes. Diversifying market risk will become less important as the growing retiree base turns its attention to longevity risk, stability of income risk, healthcare cost risk, and inflation risk. For example, an outcome investment product might focus on inflation protection with a bucket of assets that include real assets such as commodities, inflation protected bonds, and real estate. This means that, in the future, asset class lines will blur and become less important. What will be important will be a rigorous design of new products that addresses the new risks for the retired Boomers.

This is where quants will continue to excel. These new outcome-oriented products will need a solid analytical design where the relationships across asset classes can be documented, well understood, and made to work together in the most efficient manner possible. Traditional fundamental research will be less important because the goal will not be to find undervalued stocks or bonds, but instead, to focus on financial outcomes. Understanding the interrelationship across economic variables is what we quants do for a living—it is our bread and butter.

Certain Uncertainty

I wear the badge of a quant proudly. It wasn't easy to earn it, and I have to work hard to retain it. There will always be new challenges in the financial markets. I have highlighted what I believe to be a critical challenge, but there will be many others that I cannot predict. The only thing one can predict with certainty regarding the financial markets is that they are uncertain. However, this should be a cause of rejoicing for quants everywhere because it means that there will be plenty of opportunities for us to apply our skills. Economic relationships change all of the time and new ones are born every day. Uncertainty in the financial markets is another way of describing a full employment act for quants everywhere.

If I did have to strike a note of a caution, it would be not to overanalyze economic relationships. I have noticed this tendency in myself—a desire to be able to describe all economic relationships in an analytical format. Sometimes, the data just do not fit our quantitative models. This might explain why behaviorial finance has become so popular and, at times, successful at explaining economic phenomena that do not otherwise fall into a rational empirical relationship. My point is simply that not every economic observation can be distilled down into a convenient analytical format. So, be careful out there.

Chapter 10

Bjorn Flesaker

Senior Quant, Bloomberg L.P.

> Organized anarchies are organizations characterized by problematic preferences, unclear technology, and fluid participation. . . . such organizations can be viewed for some purposes as collections of choices looking for problems, issues and feelings looking for decision situations in which they might be aired, solutions looking for issues to which they might be an answer, and decision makers looking for work.[1]
>
> —from *Administrative Science Quarterly*

The authors of the Garbage Can Model studied decision making in universities, but the characterization of an organized anarchy applies quite well to the typical sell-side derivatives quant group and its immediate environment. I have personally found the Garbage Can Model to provide a useful framework for understanding and interpreting the behavior of organizations where I have worked, including, but not limited to, the importance of solutions looking for a problem, the importance of decision making opportunities, and the significant degree of randomness in actual choices made.

My familiarity with the Garbage Can Model is a direct result of having enrolled (and actually studied) at the Norwegian School of

Management following the completion of high school and military service. Is this an important part of how I became a quant? Yes and no, probably.

I took my first quant job at Merrill Lynch in October 1992, so, strictly speaking, the story should end there. I will ultimately comment on some aspects of my subsequent life, in the sense that they will help answer the more relevant question of how I came to be perceived by some people as a successful quant. I have identified half a dozen "big" decisions, along with people and circumstances surrounding them, that I will address as well. In doing so, however, I could not help recalling scenes from the movie *Sliding Doors*, in which a single minor incident does or does not happen in the opening scenes, and where the viewer is exposed to the two resulting sample paths of the main character's life. The degree to which life is random, path dependent, and highly sensitive to initial conditions is such that when you end up somewhere that you did not plan to go, it is not really possible to give an honest description of how you got there. I did not plan to become a quant. I don't think any of us did.

Growing Up

I spent my childhood in a small town in Norway with a vague sense of wanting to become a businessman when I grew up. Not that I actually pondered the distant future very much. While always doing extremely well in school, I focused most of my energy and attention on my career as a competitive cross-country skier, occasionally sidetracked by tennis, orienteering, and miscellaneous other sports. Aside from providing me with a close-knit group of friends that largely stayed out of trouble through our teenage years, I credit my experiences in sports with a lot of positive character development.

Being fairly competitive by nature, I learned how to win and lose with an appropriate degree of grace. This was especially important in an environment where I mostly competed as an individual against and alongside close friends and teammates. A big race was perhaps not quite like bonus time on Wall Street, but there are certainly similarities in terms of how you want to handle yourself. A side benefit of my relatively high rank as a cross-country skier was that I got to spend my compulsory year of military service in a sports platoon in the King's Guard around Oslo rather than being shipped off to the Russian border to endure the really long, dark, and cold winter at a latitude of 70° N.

My family is largely made up of intelligent, talented, and undereducated people. I was raised in an environment that valued educational accomplishment as a way to get a job; but I did not know many people who had a university education, let alone people working as scientists or engineers. So, after ninth grade, I exercised my right to choose a high school that was heavily focused on business studies, followed by undergraduate studies at the Norwegian School of Management. These choices were clearly influenced by the vocational roots of the schools, but in both cases their academic ambition level turned out to be fortuitously high.

Consequently, however, I am probably unique among my peers in having taken my last science course as a high school freshman. "Natural Science," a course covering physics, chemistry, and biology, was taught by a bearded Marxist whose heart was really not in his job. I did well enough in the class, but I was hardly inspired to pursue further studies in the hard sciences. That was perhaps a good thing, since the nature of my high school was such that no further science courses were offered, anyway.

Fortunately, both my high school and college offered a reasonable collection of math courses, completely liberated from physics and engineering as the default applications. Thus, my first encounter with applied calculus was in a high school business class, where we were trying to find the profit-maximizing quantity that a monopolist would offer to the market. My first exposure to vector operations involved the inner product of a list of prices and quantities. And so on . . . All in all, not a bad starting point for what I would be doing later, but I have found that whenever people try to illustrate mathematical concepts with "helpful" examples from the physical world around us, I generally do better by politely ignoring them and narrowly sticking to the concepts themselves or trying to come up with my own intuition rooted in the world of money.

The Norwegian School of Management, locally referred to as BI (short for Bedriftsøkonomisk Institutt), does not really have a counterpart in the U.S. system. It is an independent, university-level business school, with programs ranging from part-time associate-degree studies to a relatively new doctoral program. Much has changed there in the 20-odd years since I graduated, but certainly back then the academic core of BI was a four-year program of study leading to the degree of Siviløkonom. While mixing in a certain amount of math, statistics, and a required foreign language, the program got down to business (so to speak) pretty quickly.

After basic courses in accounting, finance, marketing, organizational behavior, and business law, students were required to choose a major field of study for the last two years. I actually waffled and agonized more over this particular choice than over just about any other education or career-related decision before or since. I was attracted to the glamour aspects of marketing and international business, and international business had developed a really cool final-year project involving an extended trip abroad. I cannot for the life of me recall getting much in the way of advice at this particular stage of my studies, so the decision was pretty much all mine to make. In the end, I submitted the course selection form for finance, having decided that marketing looked a bit too fluffy and that most business was "international" anyway (at least from a Norwegian perspective), so I might as well pick the area where I felt I had an edge in terms of the quantitative aspects. I had also really enjoyed the introductory finance textbook by Brealey and Meyers, which I would rank right at the top of textbooks I have ever been through.

The finance major was taught in relatively small classes, offering a genuine opportunity to get to know the professors. Despite the historical trade-school nature of BI, the finance faculty was (and still is) heavily research oriented, and they put the bar high, both in terms of the technical level of material presented and in putting a lot of original research articles on the required reading list. Practically speaking, the ambition level was probably too high for 80 percent of the class, but for me it meant getting an unusually focused exposure to modern finance research as an undergraduate student.

The program culminated in a thesis loaded with a full semester's worth of credits. After putting forth a couple of bad ideas, I was directed toward an investigation of the dividend policy of Norwegian companies, as something that had not been covered in the literature. I wrangled with data sources in paper form as well as less than user-friendly statistics software on a Prime mainframe computer, but my war stories are clearly no match for those of older colleagues brought up on assembler language and stacks of punch card. Faced with a small panel data set plagued with heteroskedasticity, I ended up downplaying the conclusions of the formal hypothesis testing and throwing in lots of pictures. This apparently worked for the readers, who deemed me as having a "mature" understanding of the econometric techniques involved.

Choosing Academics

Still imbued with the vague, albeit by now much more informed, notion of wanting to become a businessman when I grew up, I lightly penciled in a course for myself that would involve working for a couple of years in a bank and then trying to find my way to the United States for an MBA. I applied for a few jobs and actually landed a nice position as a management trainee at Bergen Bank, one of the more entrepreneurial financial institutions in Norway at the time (a banking crisis and many mergers later, it is now part of the DnB NOR financial services group).

My finance professors, however, urged me to scrap this plan and instead head directly off to study for a PhD in finance. Their general line of argument was almost certainly a bit more sophisticated, but the punch line, as far as I was concerned, was that by following their advice I could go to school instead of work, make about the same amount of money in the short run (not much), and probably much more in the long run. Several people at BI deserve credit for not just changing my mind, but also for actively helping me get into and get funding for doctoral studies. In particular, Øyvind Bøhren went out of his way to market me as a candidate to UC-Berkeley, as well as to try and teach me what academic finance research was actually all about. Starting the application process a tad late in the game, I ended up focusing my attention on Berkeley, driven in about equal measure by overall university reputation and weather.

At Berkeley, I finally met fellow students who were more intelligent and better prepared than I was. Fortunately, I had foreseen the likelihood that this would happen, so it did not come as a crushing blow to my ego, but it still took a little while getting used to. Interestingly, at least to me at the time, I felt that I had at least as strong a background in academic finance as colleagues who came from top MBA programs, but the fact that most of them had undergraduate degrees in math or engineering from high end schools meant that I felt inadequate in my quantitative preparations. So, I signed up for as many courses in math, statistics, and electrical engineering as my schedule would allow, and over the course of my first couple of years got reasonably well caught up.

The next key decision point in my academic life involved the choice of dissertation topic. I had been hanging out a lot in the math library, looking for intellectual arbitrage opportunities—specifically, for unusual or novel tools that could be put to use in analyzing financial problems. A couple of areas that I contemplated and that other people subsequently did something interesting with were nonstandard stochastic analysis and low-dimensional chaotic dynamics. Aside from falling squarely in the "solution looking for problem" camp (a natural part of life, according to the Garbage Can Model), my main hesitation with actually trying to pursue these as primary research topics was the realization that I was still pretty far from being a mathematician, and so specializing in an exotic area of applied mathematics might not be a sensible move.

In the fall of 1987, in a most timely fashion, Bob Jarrow came to Berkeley to present the first version of the Heath-Jarrow-Morton (HJM) model of interest rate dynamics and interest rate derivative pricing. Richard Grinold, who was my prethesis advisor, gave me a copy of the HJM paper a couple of weeks before the seminar and told me to dig into it. This represents some of the best academic advice I have ever received since I am not sure that I would have immediately realized the model's importance and potential for further work by myself. The rest, in some sense, is history. I really enjoyed the paper because I was struggling to understand some of the rather abstract questions in stochastic process theory that it dealt with, and I quickly decided to work on the HJM model for my dissertation. Broadly speaking, the HJM paradigm still represents the state of the art in interest rate derivatives pricing, so having been working with it from the very beginning is definitely high on my list of success factors later in life.

In my five years at Berkeley, I met a few other people of critical importance to my career path, and life in general. It might be appropriate to start with my wife, Laura Quigg, who was a fellow student in the finance PhD program. We were also both residents of International House, a wonderful institution that brings together American and foreign students in a residence rich in cross-cultural interaction. Needless to say, some of the interaction is of the romantic kind, and a long list of I-House marriages is regularly put forth as a source of pride and talking points for fund-raising. Laura occasionally reminds me that on the momentous occasion when I proposed that we move in together, I prominently used the term *economies of scale*. Apparently, the argument worked, as did the green card references when we agreed to get married a couple of years later.

Laura and I carried out some of our early courtship in the PhD student lounge, where I took on the responsibility for keeping the refrigerator stocked for our Friday afternoon happy hours. This was also one of the places where I cemented a friendship with Rich Lindsey, and where one of our Japanese colleagues, in an ongoing effort to improve his grasp of English, inquired about the meaning of "bar food" and "barf" in the same sentence.

I was assigned as a research assistant to Ehud Ronn in my second year at Berkeley. In addition to sparking a long-term friendship, this turned out to be important for a couple of reasons, one of which I will return to in due course. After having proven myself useful on a project he was doing with somebody else, he took me on board as coauthor on two papers where we investigated the short-lived inflation futures market and the design and pricing of a hypothetical variable rate mortgage, the rate of which would ratchet down with LIBOR. This helped my understanding of the research and publication process, gained me a modicum of name recognition, and altogether accomplished much of what joint student/faculty research projects are meant to do.

Interestingly, this type of research collaboration was the exception rather than the norm in the Berkeley finance department in the 1980s. The established culture of the department was very much one of swinging for the fences: One major piece of innovative research, preferably theoretical, was deemed to be worth more than any quantity of merely respectable output. This is almost certainly the way to leverage brilliant minds to move the world forward, but it did little to teach academic survival skills to PhD students, most of whom either went straight to industry positions or did so after a stint in academia. Not that I am complaining—I believe that my life as a quant has been more rewarding on many levels than a pure academic career would have been.

A small incident took place shortly after I arrived at Berkeley. I had been promised a fairly generous fellowship as part of my admission to the PhD program and, consequently, had not brought along a whole lot of spending money. I was therefore somewhat disturbed when faced with the fact that my first semester's worth of room and board was due immediately, and no fellowship monies could be disbursed until I had obtained a Social Security number, a process that could take a few weeks.

I mentioned my predicament to Hayne Leland, who had been assigned the task of advising incoming finance PhD students. After thinking about it for a moment, he responded by pulling out his checkbook and writing me a check for $5,000 to be repaid when my fellowship came through. This generous gesture had a definite impact on my outlook on life, and more than a decade later I found myself doing exactly the same thing to help out a fresh PhD whom I had hired from the University of Waterloo, whose sign-on bonus was held up in red tape, seriously impeding his hunt for an apartment in New York City (not an easy task in the best of circumstances). What comes around goes around.

Laura and I went on the academic job market as a package deal. Our first set of introductory interviews with 20+ universities took place at the FMA meetings in Boston in October 1989. This was just after the Loma Prieta earthquake, which leveled parts of our hometown of Oakland and closed down the airport indefinitely. Quick thinking and a little hustling on the phone got us rebooked on a morning flight out of Los Angeles, and we spent the night following the earthquake driving south on I-5, listening to the radio for scattered news of freeway collapses, fires, and other tragedies taking place in our backyard. If there was ever a lull in the conversation in the following interviews, we certainly had fresh material on hand for small talk, even if we were a little short on sleep.

Fortunately, the market for finance PhDs was extremely good that year, and we landed a number of joint offers. In the end, we chose to join the faculty at the University of Illinois at Urbana-Champaign. This was in part motivated by the fact that Phelim Boyle was moving there. Laura and I had befriended Phelim when he visited Berkeley the previous year, and he had invited me to speak at his derivatives conference in Waterloo, Canada, my first conference appearance with a single-author paper. In addition to being a truly wonderful person, he would also bring a certain *gravitas* to the quantitative and derivatives side of the Illinois finance faculty, which was otherwise heavily skewed toward empirical asset pricing and corporate finance.

I never did any research with Phelim, but he was a most-valued colleague in terms of providing intellectual help and support, as well as contacts around the world. And, in perhaps the singularly most fortuitous coincident in my entire career, he shared with me some research he was doing on how to tame the terrible numerical behavior of binomial trees when applied to the pricing of barrier options. Not exactly Nobel prize–level work, but it would surely be important to anybody trying to use such methods in practice.

One strand of research that I pursued at Illinois was analyzing the optimal exercise policy and valuation of options where there was uncertainty of the actual value of the underlying asset as of the time of the exercise decision. This uncertainty would sometimes lead to ex post suboptimal exercise decisions, either in the form of having exercised an option that should have expired worthless or having failed to exercise an option that with the benefit of hindsight really had been in the money.

The primary motivation for the study was to try and move the "real options" literature a little closer to reality, and my first paper in this program, which analyzed a single decision maker problem, received favorable reviews. As a direct follow-up, I started to consider multiple decision makers competing to exercise the same option, where each player had some measure of private information about the value of the underlying asset. Competing real estate developers in a small market would be an example of such a setup. What makes this interesting from an analytical point of view is the potential winner's curse inherent in the problem: a participant in the competition will tend to win when his private information is overly optimistic.

I worked out some numerical equilibrium results that I thought were mildly interesting in a single-period model, but when I tried my hand at a multiperiod version, I made a startling discovery. Looking at a model with two players and two time periods, I found that I could strictly improve the quality of the private signals received by one of the players and, as a direct result, make him strictly *worse* off. I obviously assumed I had made an error somewhere in the analysis, but after checking it repeatedly I started to believe that the result was indeed correct and that I had made a profound and original insight with potentially far-reaching implications: Better information can be bad for you! It turns out I was right, but only sort of.

When I explained my findings to my senior economics colleague and game theory expert, Charlie Kahn, he thought about it for about five seconds and uttered the phrase "prisoner's dilemma." Lo and behold, I had rediscovered a version of the most famous example in game theory. The player with better information would now rationally exercise his option early on good news, and, knowing this, his competitor would also have to act more aggressively. The equilibrium of the game would shift from the cooperative solution where both players delay exercise to get more information, to the competitive solution where either player exercises early on good news. As a result of the equilibrium exercise decisions on average being made on the basis of less information, the value of the game to either player would decrease. My fleeting thoughts of fame in the broader world of economics were crushed, and it started to sink in how slippery information is to deal with in an applied modeling context. And my attention was starting to wander elsewhere.

Heeding the Call of the Street

In the late summer of 1992, Laura and I attended the annual meeting of the European Finance Association in Lisbon. There, I got into a long discussion with my former advisor, Ehud Ronn, who was midway through a two-year leave of absence from the University of Texas to build up a fixed income quant group at Merrill Lynch. His enthusiasm for the research he was doing there and his description of how the group was having an immediate business impact presented a refreshing contrast to the more sedate pace of the academic publishing process. High-end academic finance in the early 1990s was also not all that hospitable to derivatives pricing as a research area, so the burning interest in the topic on Wall Street was definitely intriguing.

When Ehud invited me to come and present a paper to his group a few weeks later, I leapt at the opportunity. I brought along a new paper analyzing the capped stock index options that had been recently introduced on the CBOE. Heavily marketed, these exchange traded barrier options had gained rapid acceptance in the market despite a questionable design element: the barrier would only be in effect at the close of trading each day. As I showed in the paper, this would make the options behave like digital options when the underlying index level got close to the barrier, thus blowing up the option delta and making intraday hedging really difficult.

Trading volume in the options did indeed die out after the market experienced a couple of barrier-crossing events, but that may of course have been for entirely different reasons. More importantly, the presentation and accompanying interviews convinced the good people at Merrill that I was suited for a position there, and that is how I became a quant.

My first task at Merrill was to take a look at an urgent problem the nondollar swaps desk had with nothing other than the terrible numerical behavior of the binomial tree they were using to model some very large and very long-dated FX (foreign exchange) barrier options. Already knowing a solution to the problem from my discussions with Phelim Boyle, I made an immediate business impact: taming the wildly swinging risk numbers of the existing book as well as enabling even larger and more profitable pending transactions. This put me on very good terms with the largely London-based trading desk, and laid the foundation for relationships that played key roles in future career moves.

Another long-term relationship forged at Merrill was with Lane Hughston. Lane and I were brought together in a murky organizational structure, where less-compatible personalities could have caused all manners of friction and conflict. Instead, after a few months of feeling each other out, we reached the conclusion that we could work together, which started the most productive research partnership I have ever had. Over the course of the ensuing couple of years we wrote about a dozen research papers together, covering topics ranging from option pricing with transaction costs to credit default swap pricing.

A lasting legacy of the partnership was the positive interest framework (a.k.a. the Flesaker-Hughston model), where we laid out necessary and sufficient conditions to ensure positive interest rates in HJM models, and in the process developed a new martingale characterization of arbitrage-free multicurrency interest rate processes. The name recognition and network of contacts I developed through the joint research with Lane has proven invaluable in my further career, which up until now has predominantly been as a manager of quants.

I got my start as a manager after nine months in the industry when Ehud surprisingly left Merrill Lynch to return to Texas. There were no other obvious internal candidates to run our group, and I guess it was cheaper to give me a try than to go out and recruit a new head late in the year. I managed to avoid major rookie mistakes, made a couple of brilliant hires (L. Sankar and Oli Jonsson—both of whom have their own quant groups now), and was able to combine a fairly light management burden with making direct contributions to the business.

I developed a relatively relaxed (some would say lazy) management style, which would only work with high-quality employees. In retrospect, I credit much of my later success as a manager to relationships developed in those early years at Merrill. Conversely, some of my failures may have come from the inability to leverage those old relationships and experiences and perhaps to getting myself into situations calling for a more ruthless management style, which just does not fit my personality very well.

Becoming a Real Quant Again

Coming full circle, I am writing this as I am about to start a new job—as a quant! As if to demonstrate that this is very much a people business, the job offer arose from a discussion that started at an IAFE reception in honor of my old friend Phelim Boyle, where I was a guest of my long-term (but not quite as old) friend Rich Lindsey. Having perhaps burnt out a bit as a manager after a dozen years, I am relishing the opportunity to become a hands-on research quant in my new job in the R&D department at Bloomberg.

Chapter 11

Peter Jäckel

I became a quant by accident. Well, sort of. I guess I should start a tad further back.

I completed my first degree, a *Diplomphysiker*, in Germany in the spring of 1992. That summer, I went travelling all over the British Isles for about eight weeks. I visited cities, cathedrals, student unions, theatres, campsites, stone circles, forests, lakes, museums, anything really. I drove across the south from Norfolk to Land's End in Cornwall, across Wales (Have *you* been to Llanfairpwllgwyngyllgogerychwyrndrobwll-llantysiliogogogoch?), and all over Scotland. (I recommend the *Malt Whisky Trail*—very educational!) I even spent some time on the Isle of Lewis, the most northern of the larger outer Hebrides with its open peat-fire prehistoric houses and spooky Loch-side stone circles.

When I came to Oxford, I stopped by at the institute of theoretical physics and made my way straight to the head of the department to say hello. The gentleman wasn't the slightest bit fazed by my somewhat unorthodox way of asking for an appointment. He made a few minutes time to chat about the research happening at his department and about what kind of research I had done and what I was interested in. He kindly pointed me towards a certain lecturer, Dr. Tom Mullin, who was heading up a group conducting research in nonlinear dynamics and bifurcation theory at the department of Planetary, Oceanic, and Atmospheric Physics, which was part of the Clarendon Laboratory, which is technically not part of theoretical physics. Not being shy, I marched straight over and knocked on the door. As good fortune would have it, Dr. Tom Mullin also happened to be in his office, and invited me in. We had a jolly little chat at the end of which he happily offered to be my D. Phil. supervisor if I could sort out my own funding.

The English Connection

I now had a plan. I went back to Germany, and after several months of applications to various bodies for a studentship, I managed to secure some and then prepared to move my life to Oxford. I arrived in February 1993, a few days before Shrove Tuesday, and it was freezing cold. I was staying with a friend of a friend for the first few days. He introduced me to the British tradition of having pancakes on Shrove Tuesday — a hilarious event for me to watch him display his pancake high-tossing skills. I guess there is still some of it sticking to that celing today.

The next two years were a mixture of intense research work, my trying to get used to the arcane but, in a twisted kind of way, rather enjoyable customs of Oxford University, a spot of rowing for my college—I even managed to sit in the first boat for the lesser bump regatta known as *Torpids*—and making some friends.

By spring 1995, I was half way through writing up my thesis, and I submitted by the end of Trinity term 1995. What followed was a struggle to find some form of postdoc research funding. With the help of my ever so supportive D. Phil. supervisor Tom Mullin, who is today professor of physics at Manchester University, and his contacts, I secured a postdoc for a special project at the European Union's research undertaking into nuclear fusion: Joint European Torus.

My task was to analyze the observed data taken from the ongoing plasma fusion experiments to search for signatures of either genuine randomness or distinctly low-dimensional chaotic behaviour. The project was motivated by the occurrence of plasma charge disruption events.

What was happening was that whenever they increased the overall electric current, and thus particle activity, in their plasma fusion experiments beyond a threshold (which, alas, wasn't even the same everytime they tried it), the plasma current became unstable, buckled in its confining magnetic field, and the whole experiment quite literally hit a wall — the plasma current ran into the inside walls of the metal container supporting the original vacuum. These disruptions could be rather dramatic.

To put this into perspective, imagine a great big lump of metal (mainly transformer iron, but also lots of other metal parts), in total weighing about 3,000 tons, suddenly jumping a few millimetres up in the air. That's precisely what happened. These disruptive events were always accompanied by certain strange extra large magnetic fluctuations (ELMs). I was supposed to analyze whether the state of the plasma at the onset of ELMs is by its nature a low-dimensional chaotic instability, or something of fundamentally random origin.

The techniques I employed for this analysis I had learned during my D. Phil. research. The whole point of the project was that if it was to be ascertained that the instabilities are of low-dimensional chaotic nature, they could be controlled using some new techniques referred to as *Chaos Control*—in essence and theory not much different from the automatic stabilisation control you find in modern helicopters such as the Sea King. A helicopter in flight, very much unlike an ordinary aeroplane, left to itself without any hands on the control, is an inherently unstable system and will very quickly spiral wildly out of control.

It is the opposite with a trimmed plane, which is why, when you take gliding lessons, the first thing they teach you is that, if getting flustered (and no immediate threat visible such as oncoming aircraft or you are about to land), put the control stick back into the middle, take your hands off it, and relax — the plane will simply straighten itself and sit there in midair very calmly: an amazing experience! Anyway, try that with a helicopter and you might not be looking so relaxed.

London Calling

Half-way through my postdoc time at JET, I began to wonder what to do next after that contract. During my D. Phil. research, I had seen a number of very talented scientists and researchers having to go back to writing proposals for further funding, applying for contract extensions, and so on. I often felt that it is not right that a significant percentage of the intellectual capital of the country have to lead their lives on a continuous short-term employment basis. I was now facing a similar situation, and I wasn't too happy about it. So, instead of beginning to look for potential projects to apply for, and beginning to write proposals for ongoing research projects for government grant funding, I started thinking.

The only readily transferable skill I had picked up during my many years of education, as far as I could see, was that I knew a thing or two about computing. I had no knowledge of stochastic calculus, nor did I have a clue about derivatives or financial markets. I contacted some job agencies who I learned about from magazines on computing. They saw my CV and suggested that I should go for some interviews with investment banks in the City of London. I wasn't too keen on leaving the Oxfordshire countryside, but a man has to make a living, so off I went for interviews.

In the first meeting, towards the end of 1996, I had no idea what the company does I was interviewing with. I had no idea what I was interviewing for. I had no idea what a derivative is and no proper clue of what an option is. The meeting took five hours. I was both exhausted and exhilarated. The questions ranged from recursive algorithm design over Mensa-style puzzles to all areas of mathematics, in particular numerical analysis. And numerical analysis was an area I had learned a little bit about, with it first being one of my choice subjects during my physics studies in Germany, and then with it being the toolbox for all my numerical experiments I conducted during my D. Phil. research.

I loved the vibe I was getting from those people. I loved the intensity. I loved the drive in the team. I loved the commitment, enthusiasm, and dedication to the subject, albeit that I still wasn't really clear about what that was. In short, I had tasted blood. This is what I wanted to do. The energy seemed to be right up my street. It looked like I finally found a working environment where there would be no holds barred in order to accomplish.

I didn't get the job in that first interview, even though they asked me back for a second session which lasted six hours. I wasn't demotivated by that—on the contrary. I kept going for interviews with a few more places, enjoying each meeting as a free tutorial. I finally began to have a vague idea of what it was these companies were doing. That clearly helped me in subsequent interviews and, at the beginning of 1997, I had the first job offers.

The first one was from a commodity trading company's investment team. They were dead keen on all sorts of time series analysis and pattern recognition methods. They made an offer, but I had to realise that, given the extra expense of living in London, it would have meant that my standard of life would actually go down, rather than up. It really was very low for a job in finance, less than what a teacher is paid in London today.

Almost at the same time, I had some meetings with Henrik Neuhaus from Nikko Securities' European operation, Nikko Europe. He had only just joined them in order to build up a risk management group that was to take care of market risk advisory and model validation. He had one experienced investment banker with knowledge in quantitative analytics already in his team: Yasuo Kusuda. He now needed, first of all, a helping hand to build up a bit of risk-management-owned analytics in order to be ready for model validation. That's where I came in.

Nikko had around the same time finalised the hiring of Bruno Dupire and his two most experienced colleagues Antoine Savine and Guillaume Blacher. All three of them were due to join in the summer of 1997. As soon as they would start, Henrik Neuhaus anticipated that they would implement many models the company would then need to have independently validated.

Cutting One's Teeth

So, here I was in my first job in finance, and my first task was to validate the brand new and hot-off-the-press local volatility models from the inventor himself! Suffice it to say, I learned a lot in very little time. Not only did I have to learn about finite differencing solvers, forward Kolmogorov equations, Monte Carlo simulation techniques, and general derivatives replication theory quite literally in weeks rather than months, I also had to try to implement some basic building blocks of a derivatives analytics library and start thinking about how to use distributed computing in order to accomplish the envisaged full-scale hedge-test simulations with the front office models. Luckily, I had both Henrik Neuhaus and Yasuo Kusuda I could ask and learn from. I owe them both a lot for what they taught me in those first six months. Well, Bruno and his team arrived and started building models, and I started validating them.

Then there was a change of plan in the minds of senior management of Nikko Securities. Previously, it had intended to go for head-on competition with Nomura, et al. for international coverage out of Japan. All of a sudden, though, the declining share price of Nikko began to worry senior management, and a deal was struck with Citigroup: Sandy Weill, the CEO of Citigroup, agreed to buy about 25 percent of the shares of Nikko, thus injecting cash into the company and bolster the credibility of its management. In return, Nikko would close down all of its overseas derivatives operations.

For Nikko Europe, that meant redundancy for more than 500 people, yours truly included. I was 17 months into my first job in the City. So up I picked myself and went to look for another job. Things had changed for me since I entered the world of derivatives, though. I now had a little bit of experience in derivatives theory and model validation, and many houses in the City were desperate for qualified model validators, due to the regulators creating more and more pressure for in-house independent model validation to be taken seriously.

This time, it took only a couple of weeks to have two job offers, one of them being from NatWest, to start and head up a model validation group there. I was also given to understand that Riccardo Rebonato was to join some time soon. This sounded reasonably interesting, so I went for it.

All the Models in the World

The first thing I realised was that there had been no such thing as independent model validation before. It hadn't even been considered. As a consequence, the front office groups met some of my enquiring questions with a considerable amount of scepticism, and sometimes with outright suspicion. I guess it must have felt kind of weird to have the new guy snooping around.

The experience taught me a few valuable lessons, though: build your models transparently and to the highest standards so that you don't have to be afraid to be open about them; and bear in mind that you don't own the models you build for the company you work for—in fact, as quants, we own nothing, not even the paper we write on. We deliver a service to our employers. The company may have to share part of what we develop with other teams in the company and is at liberty to do so. The company may be obliged to share what we are doing with regulators. Always be aware that what we develop and implement may be subject to scrutiny.

Most quants are still not used to that: it is a common misconception that when it comes to implementation code, due diligence is not required at the same level as is expected in mathematical documents. The reality is rather different, though: mistakes in documents can be corrected comparatively easily, and require no changeover sign-off. Errors or sloppiness in implementations, however, can give rise to all sorts of problems: trading losses, overnight report calculation failures, risk limit excesses, erroneous cash payments to counterparties, late or incorrect settlement, and any other type of embarrassing situation a derivatives operation doesn't want to be seen in. Not all of the front office quants I had to deal with at NatWest were secretive about their models, though. In fact, it was rather the exception than the rule, but I learned how defensive some quants can be about their work.

Riccardo Rebonato arrived in the risk department of NatWest to become the global head of market risk, and I got involved in a whole range of interest rate modelling research. I learned a lot from him about interest rate models, most notably market models. It was a bit lucky that by that time I had gathered some experience with Monte Carlo simulations which turned out to be so crucial for the Brace-Gatarek-Musiela-Miltersen-Sandmann-Sondermann-Jamshidian world of discrete tenor models in the Heath-Jarrow-Morton framework of arbitrage-free term structure dynamics. It was a very intense period of research: swaption approximations, fast drift simulation algorithms, parametric early exercise domain optimisation, nonrecombining trees, implied skew and smile modelling, jump-diffusion dynamics, Levy processes, you name it. I can thoroughly recommend to anyone who wants to make it as a quant to spend some time in a model validation group: you get a lot more time to dedicate to analysis and learning than in the front office where there is a constant and relentless pressure for delivery of directly usable analytics.

For Future Reference

It was also during this time at NatWest that John Wiley and Sons approached me with a proposal to write a book on Monte Carlo methods in finance. At first, I wasn't keen on the idea at all because I had a hunch that it would absorb almost all of my private life for some time (which turned out to be correct).

There were two points that most likely won me over, though, to do it. The first one was the fact that I really would have appreciated having a book available on the subject when I first had to use Monte Carlo methods in derivatives pricing applications and wanted to learn about it.

The second point was the myth about Sobol' numbers not being usable in applications whose mathematical dimensionality goes beyond single digits. This myth really had me very confused at first. Consider this: here I am (back in my days at Nikko), starting my new life in finance, and I first have to learn all sorts of things about Monte Carlo simulations. I learn things, and I try things. My ingenious colleague Yasuo Kusuda points out to me the existence of *Sobol'* numbers which are described in *Numerical Recipes* and are the complete opposite of *random*. This is their key to being *better* than random numbers.

So, off I go and experiment with those. I notice that the straightfoward code in Numerical Recipes only goes to a fairly low number of dimensions: 6! Whilst they tabulate the required *primitive polynomials modulo two* up to dimension 160, the equally crucial initialisation numbers are only there up to dimension 6. Okay, I thought, I will just have to make up some more initialisation numbers up to dimension 160 following the simple recipe they give: "The starting values for this recurrence are that $M_1, \ldots, M_q$ can be arbitrary odd numbers less than $2, \ldots, 2_q$, respectively.". Arbitrary numbers? Sounds like I simply need to fill them in with random numbers that are odd and less than their individual upper limits. And that's what I do.

Then I try them. They work fine. Any dimension, well, up to 160, given that that's all the primitive polynomials I have. I had meanwhile learned from some presentation slides by Mark Broadie and

Paul Glasserman that it is always a good idea to order dimensions by importance when using low discrepancy numbers, and duly pay attention to this fact. Also I had learned from them that the Brownian bridge, for discretized Wiener process path construction, helps to tickle (nearly) the maximum benefit out of low-discrepancy numbers, while not having to use a full matrix multiplication for each path (as one would with the theoretically superior spectral construction method), which could dominate the computational effort. Alright then, I say to myself, there are a couple of minor caveats to heed when using Sobol' numbers, but that's fair enough, given that they are the equivalent to nitromethane-fuelled engines in Indy car races: you just ought to be a bit more careful than with ordinary engines!

From then on, I keep using Sobol' numbers and keep being baffled by but one thing: around the time, a whole range of publications appear that claim that Sobol' numbers dont work in dimensionalities higher than, say, a couple of dozen. I am confused — I simply don't see this problem. Since they work well for me, I want to use them for even higher dimensional problems. At that point, I have to learn about primitive polynomials modulo two, and learn how to compute more than the 160 given in Numerical Recipes.

Not being a number theoretician, I don't find the most elegant way of doing it, and thus my computations aren't as fast as I had hoped. I first find a way of computing some more using the Maple software package. I hit an upper limit with that package in terms of computation speed and memory usage, though. Thus, I write my own algorithm, and start it up. After the first power outage in the grid where I used to live at the time, I rewrite it to be restartable, and relaunch it.

Twelve months later (during which I made good progress on the book), I had computed about 8 million of them. I guess I had enough polynomials for practical use after a few days, but simply got greedy. As it happens, there is an algorithm (Alanen and Knuth 1964)—I should have known! that was later enhanced (Sugimoto 1979) that can compute these quite a bit faster—What was 12 months on a somewhat slower machine in 1999 becomes effectively an overnight calculation on nowadays' hardware using their algorithm[1].

[1]Sean Erik O'Connor. "Computing Primitive Polynomials," www.seanerikoconnor.freeservers.com/Mathematics/AbstractAlgebra/PrimitivePolynomials, 1986.

Well, I now had more polynomials than I could ever digest. And still I couldn't see any point in *not* using Sobol' numbers. There were so many suggestions to "enhance" their performance.

To date—taking into account all the extra inconvenience in applications—I still don't see the point in these "enhancements." As far as I am concerned, the myth about Sobol' numbers not working for high dimensions was just that: an unfounded rumour. I did develop a theory as to why some people think that Sobol' numbers may not work well in high dimensions, though: I found out that some theoreticians took the conditions on the initialisation numbers to mean that they could simply choose them all to be simply 1 (!). This is what I call "unit initialisation." I was surprised to find out that anyone would have thought that to be a reasonable choice. Lo and behold, when setting all initialisation numbers simply to 1, Sobol' numbers *do* perform poorly in anything more than six dimensions or so! What is intriguing about this fact is that this problem of ill initialisation had already been noticed, understood, and remedied by Sobol' himself almost as early as when he invented his number sequence in the first place. In 1976, he even published a paper with a clear description of conditions that the initialisation numbers should meet in order to be guaranteed to be useful even in high dimensions (Sobol' 1976).

To put this into context, this is the same year when Fischer Black published his generalisation of the Black-Scholes-Merton framework to options on futures contracts on commodities (Black 1976), which today is seen as the foundation for the pricing of options on underlyings in their own natural measure (vanilla interest rate swaptions, for instance, are quoted as *Black* volatilities of swap rates in their own natural annuity measure).

In other words, we are not talking about the latest findings that simply haven't been published yet, but about mathematical research that came out around the same time as Sobol' sequences themselves, and one year before Boyle's pathbreaking first publications on the application of Monte Carlo methods to generic option pricing (Boyle 1977)! I later learned from Sobol' himself that he also never would have thought of unit initialisation as meaningful, and that something like arbitrary ≈s random initialisation (within the given constraints) is much more intuitive. Well, all of this is the second reason why I felt I should write a book on Monte Carlo methods after all.

To the Front

Looking back now, I am pleased to see that Sobol's work finally receives the recognition it always deserved in my opinion. More and more practitioners and academic researchers have come to appreciate the fact that reasonable initialisation of Sobol' numbers makes them work well beyond a few dozen in dimensionality (Joe and Kuo 2003). In a nutshell, Ilya Sobol' is one of my all time heroes of applied mathematics.

All good things come to an end, and it was no different with the situation at NatWest. First, the Bank of Scotland launched a hostile takeover bid. That was quite some shock. To make matters worse, the Royal Bank of Scotland, its slightly larger cousin, followed suit with a counterbid. At that time, it would have been difficult for NatWest to fight off a single hostile takeover bid, let alone two. I really liked Derek Wanless who was the CEO back then, but he and his board were unprepared for such an audacious act. It felt like the rabbit swallowing the python. Well, they pulled it off: respect to the Scots!

The moment the Royal Bank of Scotland took over NatWest's capital markets operation, the culture changed. To many, it felt that the new senior managers, who were used to decisions being driven by credit committees, struggled somewhat with the wild wild West of the derivatives world. Many people left the firm. It was time for me to move into a front office role.

Commerzbank Securities had been trading derivatives for a small number of years when senior management decided to merge its equity and interest rate quant groups into one. They hired Olivier van Eyseren to head up the new combined group, integrating the previously entirely separate equity, interest rates, and FX analytics groups. Olivier had been a quant at Paribas and UBS and brought with him a good deal of experience in both derivatives theory and analytics library design. His vision was to build a new library from scratch that was going to have its own structured products description language, a whole host of modelling frameworks for the various asset classes as well cross-asset class hybrids, and sophisticated numerical pricing engines.

It was an exciting project, and he had put together a strong team for the job. I guess it was for most of us the first time that we had been involved in building what amounted to a whole derivatives pricing system from scratch. Everything had to be coded from the ground up: ISDA day counting and accrual conventions, holiday calendar handling, volatility quotation conventions, settlement delays, yield curve stripping, simple analytical convexity corrections, a whole host of simple option analytics (the usual suspects: baskets, barriers, etc.), number generators, multithreaded Monte Carlo simulation, variance reduction techniques, multidimensional tree solvers for diffusion-based models, general finite differencing solvers for jump-diffusion based models, multifactor Hull-White models, LIBOR market models with global calibration, Bermudan Monte Carlo techniques, serialization of any of model or product objects for possible storage or distribution, distributed valuation framework, etc. And all of this was done in a generic setup where *per se,* a *model* object would not know a priori what kind of product it is to be used for since *everything* (well, almost) was described in the in-house "Structured Products Language in Financial engineering."

I have to tip my hat to Olivier, it was a great vision, and it was accomplished, albeit at great effort. There are of course not only advantages but also disadvantages to such a generic setup of a quant library. The disadvantages aren't limited to the significantly increased effort necessary to integrate anything new. This kind of generic framework typically makes all little, simple, tasks a lot more complicated, both for developoers, and for users. Pricing up a "simple" Bermudan swaption, for instance, specified in the generic product description language based on cash flow definitions, requires a lot of initial work.

The advantages, however, become evident once anything bespoke, or slightly out of the ordinary, is to be valued: the possibilities for defining product features are seemingly endless. Since this was one of the key objectives of the library design, it was definitely a masterful achievement. I learned a lot during those three-and-a-half years, including that it is important to have the right balance between a general and a bespoke approach for any individual project.

By mid 2003, media criticisms of the CEO of Commerzbank Securities were intensifying and rumour had it that he might be planning an exit strategy in order to set up his own hedge fund. Around that time, Olivier van Eyseren left the firm to pursue a new career with Bear Stearns. My colleague Carl Seymour and I were appointed coheads of the Financial Engineering team. For the first six months, we worked even harder than before, trying to sort out all the things that may have been slightly neglected towards the end of Olivier's reign, and trying extra hard to win over senior management that what they have in their analytics library and Financial Engineering group is really better than they could ever hope for.

As our first year as coheads came to an end, though, we both felt that senior management was not recognizing the importance and key function of their quantitative analytics group to the extent that it deserved, and subsequently both Carl and I left in mid 2004. With hindsight, it is now clear that top level senior management was at least partially preoccupied with other matters: two or three months after Carl and I left, Mehmet Dalman, the then CEO of Commerzbank Securities, and all of his senior managers, jointly resigned and, indeed, went off to start up their own brand new hedge fund!

I moved on to ABN AMRO to start a quant group that is slightly smaller in overall scope but approximately the same size in headcount. We are looking after Credit, Hybrid, Inflation, and Commodity Derivative analytics. I now report to a managing director who used to be a quant himself, which makes more sense to me then the arrangement we had at Commerzbank. It is a good setup. A lot is being built from scratch, but, luckily, some of the mistakes that can be made in this process I have already been through, and that helps a lot. On the basis of the insights gained in my previous roles, it is now easier for me to have a hunch as to what ideas may work, and which ones are bound to fail. I guess, in this trade, as in many others, experience does pay off eventually.

Chapter 12

Andrew Davidson

President, Andrew Davidson & Co., Inc.

First a confession: I'm not really a quant. While I may have the resume of a quant, and I can spout quant sounding sentences, when it comes to true quantness, I fall a bit short.

Conjecture 1: If It Quacks Like a Quant . . .

My background is quantative enough. In junior high school, I was a star of the Mathletes team. I went to Harvard, where I studied math and physics. I even had a job at the Princeton Plasma Physics Laboratory working on the Tokomak Fusion reactors. I then went to the University of Chicago for business school, where I took courses given by Myron Scholes and Merton Miller, among others.

My first indication that I wasn't a quant came in college. While I was taking advanced math courses, one teacher wrote on my first math exam: "You have a serious misunderstanding, please see me."

The question was simple enough. It was something like:

Definition: A linear operator is such that $f(a + bx) = f(a) + b * f(x)$. Please indicate which of the following functions are linear:

a) $f(x) = y/(1 + x)^n$
b) $f(y) = y/(1 + x)^n$
c) $f(x_t) = \int y_t/(1 + x_t)^t dt$
d) $f(y_t) = \int y_t/(1 + x_t)^t dt$

I don't know why I could not answer that question then, but I think I had a difficult time treating a definition as instructions rather than as a philosophical issue. I believe I was trying to determine which of the equations felt linear to me. I don't think I fully understood that if you changed the word *linear* to any other word, the answer would still be the same. (I believe I now understand that *b* and *d* are linear, but *a* and *c* are not, so that present value is a linear operation on cash flow, but not on yield).

The next year, in a class on group theory another professor wrote next to my proof: "This is not mathematics." While this question has faded from my memory, I still remember the nature of my failing. I viewed groups and their elements as things that had certain relationships between them. To me, a proof was spelling out how one relationship lead to another. But, for my mathematics professor, a proof was a series of logical steps leading from definitions to lemmas to theorems. While math professors may skip some steps in a proof, there is never a logical lapse.

Nevertheless, despite my lack of aptitude for true math, I persevered and passed enough math and physics courses to graduate. Many of my courses were in mathematical physics and related to quantum mechanics. Little did I know at the time that the diffusion equations and probability operators that I was studying then would prove useful later in life.

In business school, I studied finance and statistics, subjects that I found to be quite easy with my physics background. There was so much time spent on what I thought were trivial mathematical calculations that sometimes it was hard to even find the finance. However, a few concepts have become ingrained into my thought process: systematic and unsystematic risk, construction of risk-neutral portfolios, cheapest-to-deliver. After business school I began my career, first at Exxon, then at Merrill Lynch and finally at my own firm, Andrew Davidson & Co., Inc.

While my career path has quant written all over it, as I said at the outset, I am not truly a quant. I would rather guess at the answer to a problem than work through the equations. While I studied a lot of math and physics most of my studies were of theoretical subjects, while I know a wide range of mathematical subjects, I know very little of solution techniques. I have met some great quants, and the best of them know mathematical theory as well as mathematical technique. They are consistent, logical, and diligent. (Many fine quants are profiled in this volume.) I am also not a very good computer programmer. I know a bit of FORTRAN and APL, but not enough to do anything very serious. This is the story of how I manage to survive in the world of quants.

Lemma 1: If You Don't Know Where You Are Going, Any Road Will Get You There

When I graduated from business school I was far more interested in working for a multinational corporation than on Wall Street. Along with math and physics, I had studied international economics in college. In business school, I also studied taxation and accounting. When I graduated from business school in 1983, I was offered a job in the treasurer's department at Exxon. It was a dream come true.

At the time, Exxon's treasurer's department was considered one of *the* spots in finance. Exxon managed much of its pension fund internally, including a large S&P500 index fund. It had also begun to issue its own debt, bypassing Wall Street bankers and fees. Exxon had global operations and had applied the latest thinking in project analysis using discounted cash flow methods and was analyzing and hedging the impact of currency changes on its operations. It should have been exciting. All in all, there couldn't have been a more stifling place to work.

Exxon had layers and layers of management. All decisions were made through the editing of memos. As an analyst you would write a memo on a subject, making a recommendation. (The most common opening line of a memo was, "the Treasurer's Department takes no exception to the proposal to . . . .") Your memo would be edited by your boss, who would cross out virtually everything you wrote. His version would be edited by his boss, who would cross out everything your boss wrote, and put back in half of the lines you had written in the first place. This process would continue up the ladder four or five levels. (And remember at that time most memos were typed on typewriters, not on word processors. So the secretaries were kept very busy typing and retyping the same memo.)

Once I found the opportunity to save Exxon several hundred million dollars. I was very excited; I thought I could really make a difference in the vast bureaucracy. Exxon had a multibillion-dollar tax dispute with the IRS. While the dispute was ongoing, Exxon could either pay the disputed amount or wait until the final settlement to pay. If Exxon paid now and then won, the United States would pay back the disputed amount with interest; if Exxon didn't pay, it would have to pay the disputed amount plus interest.

Exxon had two methods of analyzing projects: If it was an investment, it discounted cash flows at their cost of capital; if it was a financing, Exxon discounted cash flows at their cost of debt. Exxon was evaluating the tax dispute as an investment. If it gave the money to the United States, it would only earn at the statutory interest rate, while its cost of capital was 16 percent. According to Exxon's policies, since the outcome of the dispute was uncertain, it was an investment, not a financing.

I wrote one of those memos, explaining why Exxon should pay now and reduce its after-tax cost, since Exxon had a lower cost of debt than the federal rate for disputed tax amounts. While the litigation was uncertain, the decision to pay now or later was a financing not an investment. I was told that my analysis was correct, but that Exxon had established policies and procedures that need to be enforced around the world, and it couldn't go changing them just to save a few hundred million dollars here and there. It was time for me to go. (Years later, Exxon lost the dispute and paid several billion dollars in interest charges.)

Lemma 2: Pay No Attention to the Man behind the Curtain

Fortunately, at the same time that I was becoming frustrated at Exxon, several friends of mine had found their way to Merrill Lynch to work on mortgage-backed securities (MBS). In 1985, the mortgage market was growing rapidly, and computers were becoming more widely used on Wall Street. The market was filled with obscure conventions for quoting and trading mortgages, interest rates had been deregulated and were soaring; most of the lenders were losing money on their portfolios. My friends knew enough finance to know that mortgages had imbedded options due to the borrower's rights to prepay and to default. That is, they could call their debt or put their home to the lender. They hired me to work out the mathematics.

If I knew then what I know now, I would have realized that I wasn't the right guy for the job. Fortunately, I didn't know any better. While I could get a good start on the problem, after a while the mathematics and programming requirements were going to exceed my abilities. In mortgage research there are really only two questions we care about:

1. What is it worth?
2. How do you hedge it?

Or if you rearrange the terms there is only one question: Given a set of market conditions, what is it worth?

So I was given two problems to work on:

1. Develop an options-based model to value MBS.
2. Help the mortgage desk hedge its positions.

For the first problem, I was given a model a former employee or consultant had developed. It supposedly was a model that would value the embedded option in the MBS using a finite difference method. Given my poor programming skills and lack of training in actually solving differential equations, I never succeeded at getting the program to work.

A fundamental problem facing traders at the time was that MBS were behaving poorly. Mortgages traded at a spread to treasuries based on their average life. Therefore, it was clear that if you bought mortgages and sold treasuries (or treasury futures) you could hedge your position. Sophisticated analysts had brought Macaulay duration to the trading desk so traders would know how to hedge the positions. Unfortunately, it didn't work. As rates fell, the mortgage securities refused to rally and the hedgers lost money.

While the traders were having a hard time with hedging, the quants were not doing much better. The standard theory of the mortgage, that it was some sort of callable bond, had a major flaw. The price the MBS shouldn't be above 100. Yet market prices on some securities were around 106. It seemed neither theory nor practice made sense. Yet as we studied the market more, we realized that the market was producing prices that reflected a better understanding of option pricing than the traders or the quants.

During this time, I learned one of the most important lessons in mortgages: *How traders and other participants describe the operation of the market, and how it truly operates, are two different things.* Descriptions of market dynamics produced by traders and models should be viewed with great skepticism. For one thing, most traders talk about yields and yield spreads.

What is yield? It is the internal rate of return (IRR) of the cash flows of a project, or in this case a bond. In business school we learn when IRR should be used to evaluate a project. That is, never. Net present value is a far superior method. Yet the bond market continues to use yield and differences between yields, known as *yield spreads*, as the basis for most value comparisons. For mortgages, this problem is compounded because the cash flows are not known and in fact depend on the very yields that are being computed. Price is the most reliable market information. Much of the real work of mortgage-backed securities analysis is overcoming the barriers created by yield calculations.

Theorem 1: If It May Be True in Theory but It Won't Work in Practice, Get a Better Theory

One day as I was working on my theories, I got a call from the trading desk. Although the theories were not yet a working model, it was time to put the ideas to work. With some trepidation, I began to execute hedges, mostly shorting treasuries, and bond and note futures, and buying calls on the note future. Applying these ideas without having models developed was like changing a tire while driving on the highway. It was quite scary at times, but the pressure forced me to accelerate my understanding of MBS.

By carefully studying price data and prepayment data and through my daily activities on the trading desk, I came to several important conclusions:

- Borrowers face costs to exercise their options, which lead to above par prices for MBS.
- Borrowers have different thresholds for exercising those options, creating a phenomenon called *burnout,* whereby prepayments rise and then fall even if rates are stable or falling.
- MBS analysis requires analysis of historical prepayments. It is not possible to develop a model of MBS based solely on theoretical considerations.

This last point is probably the most crucial. Since mortgage analysis can't be done without prepayment modeling, being the best quant wasn't going to be the key to success. To succeed in MBS, you needed to know financial theory, but you also needed to study data.

Out of this work and the work of others at the major investment banking firms, the option-adjusted spread (OAS) model for mortgages was developed and refined. The first use of the model that I can recall was by the mortgage research team at Salomon Brothers, who used the model to value adjustable rate mortgages. The fundamental idea behind the OAS model is that it is possible to simulate the performance of the MBS over a wide variety of scenarios. If those scenarios are constructed properly, in line with financial theory, it is possible to value the mortgages as a spread to the discount rates.

Over the years, the process has gotten more scientific as quants have attacked the original model from all directions. The early models generally simulated one interest rate, usually the short rate, using either binomial trees or random number generators. They may or may not have been fit to the yield curve. All other rates were specified as a constant spread off the short rates. Prepayment models were simple functional forms linking interest rates, loan age, and pool factor to a monthly prepayment rate. Users of these models now expect an extraordinary degree of accuracy and consistency for these models.

In 1986, mortgage analysts were thrilled to have tools that could describe the general trends in mortgage performance. In 2006, a model is expected to differentiate value between slight variations in structure or collateral.

Despite the improvements in the models, over time I have learned that for mortgages, most models are incomplete or inaccurate. The performance of MBS is driven by borrower behavior that is extremely complex. At the same time, the mortgage instruments that have been developed have complex cash flows. The CMO market has learned to create instruments that take advantage of flaws in the methods used by investors to analyze them.

In the last few years, I have worked on the issue of why mortgages have an OAS that is not zero. This is another example of why not believing in the theory can sometimes be useful. Ever since the first option-adjusted spread models for MBS, there has been the elusive spread to explain. The idea behind OAS models is that prepayment models should capture the variability in prepayments, so once prepayments were modeled mortgages should have an OAS of zero, similar to prepayment driven assets.

However, the market never really agreed with the theory and continued to price mortgages at nonzero OAS. MBS have OAS levels that vary considerably from those of treasuries and agency debt. Not only do the spread levels vary over time, but they also vary between instruments. Although OAS should have unified the mortgage market, it just created a new set of demarcations. And as I mentioned before, it is better to trust the market than the model when describing the performance of MBS.

Recent work I have performed with a colleague, Alex Levin, who is a true quant, has shown that the OAS of mortgages is a result of the uncertainty of prepayments. We found that there are at least two types of uncertainty that are priced into MBS, refinancing uncertainty and turnover uncertainty. We find that the market is priced to be biased against more efficient refinancing and slower turnover. Thus, the pricing of the mortgage market reflects a type of risk neutrality, where the market does not price to expected cash flows but to risk-adjusted cash flows.

This work brings us closer to understanding MBS, but it is not the end. There are still more anomalies and discrepancies to explore.

Theorem 2: To Thine Own Self Be True

Most quants with any sense move on from mortgages after a while. They find that the problems are too elusive. Analysis of mortgages involves ad hoc assumptions that are continually changing. There is no grand theory that captures everything; every theoretical advance is met by an obstacle in practice. Instead of having a unified model to solve problems, mortgage analysts must be content with keeping a tool bag of ideas and applying those ideas to problems as they arise, knowing that there is some important ingredient or market innovation that will make their conclusion wrong.

So either because I am not truly a quant, or because I don't have any sense, I have remained in the mortgage market. For me the challenge is not just in gaining a better mathematical understanding of the market, but also in understanding the complex interactions between borrowers, lenders, and investors. Because of its size and scope, the mortgage market touches all aspects of the financial markets and the economy. By striving to understand the workings of the mortgage market, I have also gained a greater understanding of the workings of the economy.

Although I am not the greatest mathematician, this weakness may be a strength in a product where the equations never work out quite right. A clear conceptual picture often is a better solution than a mathematical model that misses a key component. It is a market where someone who isn't really a quant can have a great career.

Chapter 13

Andrew B. Weisman

Managing Director, Merrill Lynch

My quant career began in an unassuming manner. As an undergraduate at Columbia College, I was steered in the direction of the classics: Plato to Popper, Beowulf to Virginia Woolf, and, oddly enough, swimming lessons. There was, of course, the small matter of declaring a major designed to provide me with an entrée to graduate school, or some form of Wall Street training program. I selected economics and philosophy, thereby gaining a foundation in econometrics, statistics, calculus, and formal logic. I felt this would make a fine compliment to my vast repertoire of arcane highbrow chit chat. Grad school followed: first at Columbia's School of International Affairs studying international economics and business (more calculus, statistics, and econometrics) and then Columbia's Graduate School of Business in the PhD program in money and financial markets. Interestingly enough, Columbia's doctoral program has a three-language minimum:

English, some other language chosen to facilitate the reading of academic literature, and finally mathematics. Most of the enlistees came prequalified—especially those emanating from India in general and the India Institute of Technology (IIT) in particular. I must admit that on many occasions I felt distinctly outclassed. Nonetheless, I soldiered on and even managed to obtain the top score in my year for the department's comprehensive exam. It was a bit of a surprise for all those involved and a real testament to allowing the power of anxiety to be a positive force in your life.

Econometric Voodoo

Upon completion of my course work and comprehensive exams, I began a consulting relationship with the Foreign Exchange Department of Shearson/Lehman Brothers. The basic terms of the relationship were that I would deliver econometric voodoo, through the use of such forecasting tools as vector auto-regression (VAR), auto-regressive integrated moving average models (ARIMA), and Kalman filters, for the purpose of forecasting movements in the foreign exchange markets, and they would teach me the dark art of speculation. There was, of course, a vast array of literature being produced at the time (mid-1980s) by the giants in the field of rational expectations econometrics such as Robert Lucas and Thomas Sargent, which clearly demonstrated through the use of well-structured tests of market rationality and neutrality that such tools were likely to prove useless. My feeling was, however, "Why let that get in the way of a good time?"

If I could just cleverly restructure some of the independent variables, focus the research on a highly granular timeframe, use a complex enough model—then who knows? Let's turn up the juice and see what shakes loose. The firm was, of course, eminently capable of delivering on its end of the bargain. I, on the other hand, muddled along, testing model after model to little effect, all the while trying to make sense of my surroundings. I was to discover that graduate school had prepared me in a decidedly tangential manner for what I was to encounter. The foreign exchange markets, which I had studied extensively in the context of purchasing power parity, uncovered interest rate parity, and geometric

Brownian motion, were almost unrecognizable to me. I frequently felt lost in a world of rapid-fire economic and political developments, blaring broker boxes, and arcane market nomenclature. It was as if I had embarked on a career as a professional pugilist armed with an extensive knowledge of anatomy and the basic instruction set that I must strike my opponent firmly about the head and abdomen while avoiding same. It was some time before the observed phenomena were to come into focus.

After a year at Shearson, my boss transferred to Bankers Trust and invited me to tag along. It was there that I had my first epiphany as a quant researcher; it consisted of the fundamental realization that most of the trading that occurs on an interbank trading desk takes place on a timescale that, for the most part, defies fundamental economic analysis. In truth, longer-term market convictions based on well-structured, defensible analysis are apt to get in the way. Such speculative activity is more akin to playing Pong.[1] It is primarily a reflexive activity, with the trader operating under no imperative to distinguish Margaret Thatcher from Terry Hatcher. In many cases, doing so would have been a tall order.

Trading for Fun and Profit

Armed with this knowledge, I set about creating my first reasonably profitable trading model. The basic idea was to conceive of a trader as a mindless maximum likelihood estimator. All trades were thought of as discrete Poisson arrival shocks to the market, separated into two processes: those that forced the market up and those that forced it down. Essentially, when the up-moves dominated, one was long, when the down-moves dominated, one was short. The model was relatively simple in conception, somewhat complex in design, and woefully intricate in its implementation. It was the work product of three individuals who might just as well have been playing Dungeons and Dragons. In the model's final form, it would decide its own trades, call one of us on the telephone, verbally instruct us (with a voice that sounded not unlike a Swedish Isaac Hayes), and await our response in order to be certain that we had understood what was required of us. It is worth noting that the automation of this model through the use of digital feed drivers and voice synthesis technology was both ahead of its time and wholly gratuitous.

In retrospect, it was an immensely circuitous method for capturing serial correlation in the foreign exchange markets. After the model's completion I took the time to read Richard Sweeney's paper titled "Beating the Foreign Exchange Markets,"[2] where I was to discover that (and I paraphrase) virtually any simple filter-rule model (essentially buying x percent above the local high and selling x percent below the local low; randomly varying the x to get more models) is capable of dominating buy-and-hold hedging strategies, such as hedging through the use of options, forwards, or futures over virtually any five-year period. The basic reason for this effect was not that filter rule models are somehow magically efficient, but rather that the underlying exchange rates exhibit significant serial correlation. The sad truth is that the anomaly itself was probably the source of 90 percent of the value associated with our trading activities; a point that was initially lost on me, but ultimately proved central to my research and investing world view. A second point, which was not lost on me, is that the market tends to place a significant premium on complexity. Basically, one should never explain a simple concept or procedure in simple terms, or it will be robbed of its marketability. Keep the curtain firmly in place and crank the devil out of your thunder machine.[3]

Tools of the Trade

My time spent at Bankers Trust was extraordinary. I was provided access to the excessive resources of a highly profitable foreign exchange trading operation: virtually unlimited trading lines, rudimentary, if not honor-based, risk management, experienced wily investors and a lax regulatory environment (or so it was thought).[4] This was the *phlogiston*[5] of legend. I can neither confirm nor deny having witnessed anything . . .

What can be confirmed is that possessing quantitative technical skills clearly accelerated my admission into the upper echelons of risk-taking. On the one hand, I was spared much of the hazing that was typically associated with joining this club. On the other hand, I was subject to certain mild forms of prejudice. Generally, too obvious an academic predisposition can result in the perception that one is incapable of making difficult, time-sensitive decisions. This perception, if allowed to persist, can limit upward mobility. One has to be wary of the *silicon ceiling*. Additionally, one should be careful not to reveal the full range of one's technical expertise, as one might become woefully sidelined into a technical support role. Trust me; it pays more to be a big picture guy. While I have nothing but the most profound respect for the technically adept, and seek to advance them whenever possible, I am very much in the minority. The majority, especially those successfully operating within large financial institutions, are more appropriately considered outside the boundaries of a meritocratic framework. Rather, one should consider the pack-dynamical framework espoused by Cesar Milan, the renowned National Geographic Channel *Dog Whisperer*. In short, learn to be calm-assertive and develop your scent-marking and dominance behavior patterns.[6]

My years at Bankers Trust had several additional significant consequences:

1. *A heightened sensitivity (or aversion) to any research produced at too great a distance from the subject matter.* While simultaneously functioning as a research assistant at Columbia's Graduate School of Business, I encountered a piece of research that had uncovered an interesting feature of the dynamics of the European Monetary System (EMS). Basically, whenever the U.S. dollar (USD) strengthened versus the Deutschemark (DM), there was a tendency for the DM to weaken against other EMS currencies. Similarly, whenever the USD weakened versus the DM, there was a tendency for the DM to strengthen versus other EMS currencies. The author of the research hypothesized that such effects were due to the disproportionate holdings of the DM in the portfolios of currency speculators. This was, however, not the actual cause of this phenomenon. I knew this because I sat next to the actual cause (hereafter referred to as AC). AC had, in fact, manufactured this effect, elevated it to a true art form, and developed his own nomenclature; it was called the *Bop Play*. Whenever AC observed the USD strengthen significantly, he would sell several "yards"[7] of DM in as boisterous and attention-grabbing manner as possible—like it was the end of the world. Having caused the market's bid-side to

evaporate, he would then square up, typically at a significant profit. The bottom line here is that while the researcher had hypothesized an explanation that was consistent with the observed events it was not correct; producing a classic "*post hoc, ergo propter hoc*"[8] error. There was no other real imperative for this effect to occur. The bottom line is that it's not a bad idea to get to know what you're analyzing.

2. *A deep and abiding appreciation for the mechanisms that can be brought to bear to advance the cause of investment performance without actually performing any better.* Such techniques were ultimately summarized in a paper submitted to the *Journal of Portfolio Management* under the title "Informationless Investing and yadda yadda yadda . . ." for which I was awarded both the Bernstein Fabozzi/Jacobs Levy Award (which was truly appreciated), and the undying enmity of vast hordes of traders who wish I had kept my trap shut (which was not as grand a distinction).

The basic performance enhancement tool kit consists of: (1) smoothing; (2) selling volatility (i.e., taking in premium while assuming the potentially significant risk of a low probability event); and (3) doubling up when you're wrong.

Smoothing

I learned of this activity while observing the firm's long-dated forward traders. This group was tasked with making a market in, for example, the Japanese yen versus the U.S. dollar for delivery 10 years forward. Suffice it to say, this was an over-the-counter market with a bid/ask spread as wide as the great outdoors, and no basic mechanism to verify the appropriateness of an end-of-period price. If the desk could find a large corporate customer to deal on the bank's pleasingly plump bid/offer spread, then the profits could be "fed in" to the firm's P&L fairly consistently over a long period of time. Conversely, if the market moved dramatically against the desk after such a transaction, the traders needed to acknowledge only a limited portion of the actual loss that had occurred. The primary benefits of this flexibility were dramatic decreases in both reported volatility and correlation. The downside was that periodically, if the market moved too significantly, the trader's mark could deviate too obviously from an appropriate level, and they could get called out. The effect of such a credibility breach is typically a forced, sudden liquidation of a potentially large and illiquid position that frequently results in both an immediate loss that takes you, if you're lucky, back to fair value but typically well beyond. Such events are periodically observed in the hedge fund world. It is worth noting that adjustments suggested in the academic literature are, in the author's opinion,[9] woefully inadequate as a corrective response to the observed serial correlation that is induced by smoothing. Such corrections explicitly ignore the most frightening, career defining, effect of smoothing; the big honking loss associated with having to sell a boatload of toxic securities in a short period of time when all your erstwhile buddies know what you're trying to do. Subsequently, smoothing is more appropriately thought of as a short sale of a barrier option, with the barrier thought of as some predefined credibility threshold characterizing the difference between where a trader has marked his book and where he could actually liquidate, and the payoff is the sudden liquidation penalty.[10]

Selling Volatility

Selling *vol* has many incarnations. I learned of it originally in the context of the *carry play*. In currency space the carry play consists of borrowing (selling) in low-yield currencies and lending (buying) in high-yield currencies. When the European Monetary System existed, such trades were extremely lucrative. One needed only the most rudimentary skills to borrow the DM at 4.5 percent and lend the lira at 14 percent in order to appear a suave international investing sophisticate. The whole thing appears to have been part of a social welfare program sponsored by European central banks; part of their "no trader left behind" program. There was, of course, the periodic risk of losing your shirt during one of the relatively infrequent currency realignment events—the weak currencies would periodically be revalued to substantially lower levels, resulting in a significant pounding if one failed to read the tea leaves and exit the trade.

This dynamic, known variously as the *peso problem*, *leptokurtosis*, and *picking up nickels in front of a steam roller*, is relatively widespread in the investing community. The truth is that it persists not because traders are unaware of how the story will end, but rather because the story can be rewritten again and again without having to give back the Pulitzer Prize; a classic agency problem if ever one there was. From a technical standpoint, short vol trades create ample opportunities for inexperienced quants to produce diametrically incorrect answers.[11]

Generally, performance data are conditioned on a manager not having experienced the key low-probability event; if they had then they would be out of business.[12] Subsequently, the event that would have provided the texture necessary to elicit an appropriate estimate of volatility is almost, by definition, never available—an odd form of financial *Heisenberg's uncertainty*.[13] This is one example of just why quantitative financial research is so darn difficult; it involves rational (or at least thinking) participants who are not typically bound by the persistent laws of the physical world.

Doubling Up

The most colorful episode of doubling up behavior involved the notorious trader Nick Leeson, who was instrumental in bringing down the House of Baring.[14] By all accounts, he reacted in a classic manner to the OPM (other people's money) agency problem; he had masterfully engineered a career-ending loss of "X" and concluded that there was no additional cost in losing "2X" (since it was not his money) and there was a substantial cost associated with not recouping the loss before it was detected. It is worth noting that doubling-up strategies (or some milder variations thereof) are fairly common in the trading/hedge fund community. To detect the use of such strategies, one need only incorporate an *equity scaling* (increasing or decreasing exposure as a function of profitability) variable into your factor model; when overall exposure "loads" in a negatively correlated fashion with profitability, you've found what you're trying to avoid.

Lessons Learned

After more than a decade as an FX trader, with its concomitant 100-hour work week, I decided it was time for a change. In the mid-1990s I managed to successfully recast myself as a hedge fund "expert,"[15] first as head of the structured products group for Yamaichi, and then as the chief investment officer for Nikko Securities International. As fate would have it, I arrived at Nikko just six months in advance of the most interesting (in the Japanese sense of the word, meaning really horrible) period in the development of the hedge fund industry—the summer of 1998. This was the time of the "perfect storm" when all the smoothing, short vol, and double-up folks were taken out to the woodshed and summarily shot. There was, however, a silver lining; this period afforded ample opportunity for reflection[16] and ultimately provided the impetus for the development and implementation of a whole host of advances in the areas of portfolio construction and risk management.[17]

The primary lesson I learned from the events of this time was the need to cast a skeptical eye on any quantitative technique that had the potential to produce a *flag pole* solution—namely, a solution that was too finely calibrated to function out of sample. In my experience the two primary culprits were mean variance optimization (MVO), and value at risk (VaR). MVO in all likelihood contributed mightily to the scale of the disruptions that occurred that year. The use of this technique, in combination with the informationless performance-enhancement techniques that are frequently employed (knowingly or unknowingly) by professional investors, tends to produce portfolios that are optimized to produce maximal future period losses; systematically denigrated from a liquidity standpoint; and actively inclusive of managers that make use of money management strategies that imply catastrophic losses of capital. It is my firm belief that unless Harry Markowitz had a truly ironic sense of humor, this was not his intent. Since this time, savvy investors have begun to employ several compensating techniques in order to accommodate the peculiarities of hedge fund data, including conditional value at risk (CVaR) in recognition of the distinctly nonsymmetric, left-tail-skewed, reality of hedge fund investing;[18] resampled optimization in recognition of the frailties of error estimation and the fact that life is, sadly, out of sample;[19] and double secret probation mechanisms for developing and incorporating forward-looking views on return, volatility, and correlation.

The second culprit VaR has been used, and is in my experience still used, to substantiate the assumption of risk that, in a qualitative framework, would be unacceptable. The most recent example that I observed (a.k.a. lost money as a result of) involved a commodity manager who went home on a Friday afternoon with a reported daily VaR of approximately 5 percent. The portfolio opened the following Monday with a measured daily VaR of 50 percent due to the arrival of Hurricane Katrina. A qualitative analysis of the structure of the portfolio would have immediately triggered alarm bells, irrespective of the sanguine musings of the VaR engine. It is worth emphasizing that quantitative measures of risk, until further notice, should not be viewed as a strictly cardinal (absolute) expression of risk, but rather as an ordinal measure; my VaR is more than it was a week ago, less than it was three weeks ago, but what "it" is, is not truly knowable.[20]

Incidentally, there was a second, albeit important, lesson that I learned in the aftermath of these events—don't talk to reporters. During the course of a conversation with a reporter from *Institutional Investor* (II), where I was summarizing some of these opinions, I off-handedly commented that: "As hedge funds find their way into the investment portfolios of 'Joe Lunch Box' (JLB) there will be an associated migration of higher-order statistical concepts. Most people know what an average is. A small but significant number know what a standard deviation is. However, when JLB starts earning 1 percent a month for 48 months in a row and then gets clocked for 30 percent, he's going to learn all about kurtosis. At the moment, though, most people think kurtosis is some form of skin disease." These comments convulsed the reporter, who promptly sent a photographer over to my office to snap a less than flattering image of me. It was subsequently published as a full-page spread in II with the byline "Living with Kurtosis." It was several months before strangers were bold enough to shake my potentially disease-ridden hand.

As I move forward into the latest chapter of my life as a quant on Wall Street, I do so with the emboldening realization that senior Wall Street management has finally come to understand that investing is an intellectual arms race. You really need the involvement of those who find joy in the inner workings of the genetic algorithm and the Gaussian copula. Besides, quants tend to accent the workplace in the most positive and entertaining ways; and there is no shame in enjoying your life.[21]

Chapter 14

Clifford S. Asness[1,2]

Managing and Founding Principal, AQR Capital Management, LLC

For most of college, I intended to be a lawyer. I think you would be hard pressed to find a profession further from quant finance. I intended this career for the same reasons most people intend such things. I hadn't really thought about it that hard, but my family was full of lawyers and I liked the 1970s show *The Paper Chase*. Even though law was my intended destination, I was studying for two degrees as an undergraduate, one in business and one in computer engineering.[3] Some time around junior year, two things happened. One, I started working for a few Wharton professors as a research assistant and found the material and process fascinating. Two, my father, a highly talented lawyer who did not understand why anyone, including himself, would voluntarily become a lawyer, wondered out loud why I would join his guild when I was "good with numbers."

Rather abruptly, but fortuitously, I canceled my date with the LSATs and scheduled the GMATs (not an apocryphal story) with a plan to pursue a PhD in finance. If that sounds like a quick change and implies I can be easily influenced, you have understood my meaning perfectly. I was accepted at several schools, with it ultimately coming down to the University of Chicago versus Stanford, each offering the same scholarship package. Chicago had it in its budget to fly me out for a visit but Stanford did not, and having little cash at the time I only visited Chicago. It was a beautiful spring day and, in the ultimate bait and switch, I've long claimed that I am the only one ever to choose the University of Chicago over Stanford based on the weather.

Chicago

After arriving at the University of Chicago, I ended up in my second year as Gene Fama and Ken French's teaching and research assistant. This was a good thing to be at Chicago's GSB (Graduate School of Business). I was lucky enough to be there smack dab in the middle of their justly famous research on value investing. When it came time for me to write a dissertation, I thought of extending their work, but quickly encountered the fact that Chicago is different from other places. My sense is that at many schools, students of famous professors pursuing a new line of research get to work on a piece of the puzzle. Kind of like "Cliff, you go study value investing in Japan." You still would have to do a great job, but you'd sort of be part of a research team. Not so at Chicago. In the ensuing years I have often referred to Fama and French as great guys, but "scorched earth researchers." Meaning, if they didn't already do it in their current paper, they were clearly working on every possible extension in their already-half-done next papers.

So, it clearly called for another angle.

That angle ended up being momentum investing. My dissertation included some other things, but was most focused on studying the puzzling trait of winning and losing stocks to statistically continue these trends.[4] In fact, one of my major contributions was showing that the short-term reversal strategy (see the whining in previous footnote) needed to be accounted for in order to truly see the power behind momentum investing. I was not the sole discoverer of momentum. Wall Street had looked at momentum for years, and Jegadeesh and Titman generally get, and deserve, the most credit for studying it first in modern academia. However, I was certainly early and made some neat refinements.

After just starting my dissertation at Chicago I got an offer to work for Goldman Sachs for what was then planned to be a year away from academia.[5] The offer was from the fixed-income group at the asset management division (GSAM). GSAM was only a few years old at the time, not the powerhouse it is today, and thus the opportunities were great. I was originally hired to be a model-building quant. However, after doing some research for the group as a consultant and spending a summer there, I quickly became full-time for a trial year, splitting my time between quant work and portfolio management.

I got my start specializing in mortgage-backed securities, and I'm fond of saying it was a great baptism, as every investing disaster that can befall you happens to some variety of a mortgage-back in reasonably short order (interest rate risk, default risk, and maybe most important, "oh my God, who knew I wrote options" risk). So if you can do well through these events, you've learned a lot.

A Big Decision

After about a year at GSAM, I was still undecided about my future. I was making good progress on my dissertation at night and on the weekends, but was still divided between choosing life as a Goldman Sachs portfolio manager or as an academic.[6] Then, I got lucky.

I will pause for a moment to stress the importance of luck in one's life path, which is often downplayed in autobiographical work. This isn't particularly useful advice (go get lucky!), and I do not mean it fatalistically, as hard work and courage really can put you in the right place to get lucky, but especially for an empiricist, to deny the role of fortune in terms of the highly path-dependent nature of life is just excessively egotistical.

My luck in this case was when someone at the West Coast asset manager PIMCO read my first *Journal of Portfolio Management* paper (the page turner titled "OAS Models, Expected Returns, and a Steep Yield

Curve" (1993)) and liked it. After some discussions, PIMCO offered me a job as head of a new quantitative area. Their idea was very open ended, but essentially it was to start a centralized group to produce quant models and tools. I told my bosses[7] at GSAM I was thinking of doing this, and they immediately replied, "We are thinking of starting a similar group; why don't you do it here?" I know what you're thinking, but I'm pretty sure they were telling the truth, as they had already interviewed someone.

After some deliberation I decided to stay at GSAM and start this new group. At that time it was called the Quantitative Research Group, even though we directly managed portfolios to distinguish it from GSAM's already-existing and successful Quantitative Equity Group run by Bob Jones and his chief lieutenant Kent Clark, both today Goldman partners. If you think that was confusing, you are right.[8] I told my contact at PIMCO about my decision and how, in an otherwise tie (two great firms, two similar opportunities), it came down to staying near where my family lived. He, in a good-natured way, said "So you're taking the worse job because you're a mama's boy, huh?" I like people who say things like that![9]

So, I started a group at GSAM whose mandate was summarized as "let's see if we can use these academic findings to make clients money." Our original business plan specified building models to forecast returns on many assets and asset classes, and noted that we could use these models not only to aid portfolio managers in making their decisions, but also to directly run client money. It also noted that if we created good models, they could be used to generate "absolute returns" in a hedge fund product, or relative returns in a constrained traditional long-only product, and we planned to do both from the get go.[10]

But, I'm getting ahead of myself. The group's first assignment was not to directly manage money, nor did it directly involve quantitative stock selection (on which I had written my dissertation), nor was it related to fixed income (which was my prior day job). No, any of those would have been too comfortable. Our first task was to try to help a traditional global equity team (fundamental stock pickers who visited management and ran a fairly concentrated portfolio) that had shown ability in pure stock selection, but had a pretty tough run with their country decisions. Thus, we got asked if we could build a tool to help make relative equity country-level decisions. This was 1994 (a tough year on Wall Street) and we were a brand new group that had accomplished nothing so far, so the obvious immediate answer that sprang from our lips was, "Of course we can help with that!" The next step was to close our door and try to figure out how we were going to do it. By this time we were a team of four—two friends had joined me from the University of Chicago's PhD program, John Liew and Ross Stevens, and a fourth, Brian Hurst, joined us from Wharton's undergrad program.

Luckily, the leap was important, but not that difficult. We quickly realized that if the value and momentum anomalies we had studied in academia for U.S. stock selection sprung from sustainable effects (risk premia and/or investor biases), then a strong working hypothesis was they would work for other decisions. We quickly realized that we could treat countries as portfolios of individual stocks. We could then buy (or overweight) the cheaper ones (France's price-to-book, for instance, is the market capitalization of all French stocks divided by the summed book value of all French stocks; if France is selling for a 1.0 and Germany a 2.0 then Germany is relatively expensive[11]) with better momentum and sell (or underweight) the expensive countries with bad momentum. It turned out that the value and momentum strategy for picking countries performed quite well.[12]

The practical upshot was we were successful in our first task as a new group, but I still think the theoretical implication was more important. One of the big problems with quantitative work is data mining. (I will argue here that it's not just quants who suffer from this.) That is, finding things that have "worked" in the past just because you looked at a lot of things and found some random coincidental effect.[13]

The best cure for data mining is an out-of-sample test (i.e., trying it somewhere else or over a different time period). Now, 15 or so years since the early 1990s research on U.S. individual stock data, we have a nice out-of-sample test (the last 15 years!), but that's not something for which we could wait. Showing that the same effects held for very different investment decisions, like choosing countries, was in reality a wonderful out-of-sample test of the power of value and momentum. I have often said that if we only traded the original individual U.S. stock strategy, we'd still be thrilled to know value and momentum "worked" for many other investment decisions, as it makes the likelihood that the

U.S. stock results were random luck far smaller. We, rather stupidly, wrote up this result in the *Journal of Portfolio Management* under the title "Parallels Between the Cross-sectional Predictability of Stock and Country Returns." Perhaps this is hypocrisy, but similar publication of more current results would be a capital offense at AQR!

In rapid succession, we built value and momentum models for other investment decisions. In addition to our U.S. stock selection and equity country selection processes, we quickly added models to pick stocks in the United Kingdom and Japan, and models to pick amongst the world's bond and currency markets (using direct analogies to the value and momentum measures we used for stocks and stock markets).

In a key lucky break, Goldman Sachs decided (at our prodding) to seed with partner capital a very aggressive market-neutral hedge fund utilizing our new investment process.[14] Although we had very strong results in general across many products (both long only and absolute return), over the next few years our results for this hedge fund were off the charts. These results were not just great, but much better than our own backtests, a key sign you're getting at least somewhat lucky as an iron-clad rule is to expect results worse than your backtest. Don't get me wrong, I think we created some great models, but getting a lucky draw on top of a great model is a pretty wonderful thing to happen early in your career. (As they say in the novel *Dune*, "Beginnings are delicate times.")[15] A few years down the road, we were managing $7 billion, about $6 billion in long-only assets, and close to a billion in hedge fund assets, all with strong-to-stellar results. That's when the siren song of launching our own firm got too strong to resist.

On Our Own

We decided to leave Goldman Sachs at the end of 1997. By this time, it was definitely "we," as I made this decision along with my current partners David Kabiller, Bob Krail, and John Liew.[16] This decision was based partially on simple ambition and greed (anyone who starts a hedge fund and doesn't admit this motivation probably shouldn't be trusted with your money), but also because our responsibilities at Goldman were too broad. Besides running $7 billion in client assets, we were also the general quant group for GSAM, doing efficient frontiers if the marketing group needed them, asset liability analysis if a client requested it, and many other tasks that were not directly about research and portfolio management. That was not consistent with our desire to focus solely on managing client money. We started our firm AQR Capital Management, LLC (Applied Quantitative Research) in early 1998 for all these reasons.

We launched our first hedge fund in August of 1998 with $1 billion of new capital, the largest standing start we had heard of for a hedge fund at that time. Why did we launch with a hedge fund when we had run both traditional and hedge fund assets at Goldman Sachs? Well, there's a paradox to this industry. To attract traditional assets, you need a five-year track record, some gray hair on your team, and it does not hurt to have the solidity of a Goldman Sachs behind you. To attract hedge fund assets you only need to be 29 years old and say that you are closing. Of course I'm being too flip. For starting either one a great track record, pedigree, and investment process are crucial. However, the strange difference in difficulty between starting a hedge fund and a traditional firm is quite real.

So we launched with a billion of hedge fund assets in August 1998, and today have about $35 billion in assets, about $9 billion in hedge funds, and about $26 billion in long-only beat-the-benchmark mandates. Describing it that way makes it sound like *way* too smooth a path. I am fond of describing our first 20 months as "We launched with 1 billion, and then through diligence, hard work, and some astute market calls, we turned that into 400 million in under two years." Then I wait for the realization that this is far less! An attentive reader might note that we launched in August 1998. Because the massive quantitative (I will dispute that characterization later, but for now it works) hedge fund Long-Term Capital (LTCM) blew up that August, that surely accounted for our problems.[17] That is a perfectly reasonable inference, but incorrect. August of 1998 was actually a good month for us. Our problems were the next 19 months, as the end of the LTCM crisis pretty much kicked off the great technology/Internet bubble. This was not a good time for us. The title of the next section could very well be altered to "How I Barely Stayed a Quant."

Basically, in a nutshell, for our first year and a half, momentum was a good strategy, but value (for picking stocks and stock markets) was disastrously bad. Given our process incorporates both, we didn't have quite as tough a period as pure value, but it was pretty tough! Accepting that tough times will happen is part of the job, but when it happens in your first year and a half in business, and you are taking 20 percent annual volatility, and you are truly market neutral in a time of markets soaring (so it's not just losing, it's posting negatives relative to others soaring!), I assure you it's not pretty.

We fought very hard to preserve our business. We made the case time and again to clients in private presentations that our drawdown was statistically not that shocking a result given our volatility and market-neutrality (we had a triple-digit gross return year at Goldman), and more importantly, that we thought we were right and due to make a lot of money when things corrected. I also made the case publicly in a series of bitter, sarcastic presentations and papers, all sharing a common title "Bubble Logic." Please note I use the words *bitter* and *sarcastic* in their traditional complimentary fashion.[18] After many publications over the years, "Bubble Logic" is still my favorite work. It is rare you get to take an extreme stance on a first-order issue to the world's wealth, get it right, and try to be entertaining while doing it. Ironically, it is one of the few pieces I never got published, as I wanted to make it into a book. It was about 50 single-spaced pages, too short for a book, but too long for an article, and as the bubble started to come down and we started to do well I tried updating it a few times, but "get out now" is much more interesting than "you should have gotten out a year ago," so eventually I relented and moved on.

In addition to making our case, we did make some changes in both our fund and our business. We researched and implemented new strategies and launched some new funds and long-only business lines. Perhaps most important to survival and most disappointing in retrospect, we cut our original fund's target volatility from 17 to 20 percent to 12 to 14 percent. Truth be told, we actually cut by more like half as we moved to some more conservative, and I think more accurate, assumptions about volatility and correlation amongst strategies. Had we not lowered our risk, it's questionable whether we would've survived, but it's not questionable that had we survived, we would have made clients a lot more than the already big gains we produced over the next few years![19] Most importantly though, we stuck with our basic process, and did not waver on our strong belief that our value strategy was epically attractive, and in fact we increased the proportion of our risk (not our total risk) that came from our value-based stock selection bet. This, needless to say, worked out quite well over the next few years.

Let me pause for a blatant suck-up paragraph (it may be sucking up, but it is also true). Most of the $1 billion we raised was locked up for only one year, and our drawdown hit bottom at 19 months long. We could not have stuck with anything if we did not have a world-class set of clients who by-and-large believed in us, believed the world was crazy (or at the least that it paid to have a hedge on the possibility the world was crazy). In some cases, our clients actually added to their investment with us near the bottom. We offered existing clients the high-water mark of departing clients. This meant that if they invested new money with us, we wouldn't take a performance fee until we had made back our large since-inception losses, even though those losses had occurred for other departing investors. At even the most optimistic of scenarios, with the power of compounding working against us, this meant we would be working with no hope of a performance fee for at least a few years. While we fully expected this offer to pay off for us, we also felt it was a nice gesture to the clients that had stuck with us through the hard times. Of course, even with this enticement it still required great courage on our clients' part to participate, and I will be forever grateful for having clients who believed in us through this very bleak time. Thank you.

Finally, the worst of times in your life can sometimes end up as the times you look back on with the most pride. So it is with the bubble that began in earnest nearly to the day we started AQR. I'm proud of sticking with and upping our bet on value while many caved to the mob. I'm proud of how hard, and how successfully, we fought to keep our clients. I'm very proud of how we retained our employees and kept our partnership (if not my sanity at all times) intact. Most generally, we kept alive and ultimately made flourish, a hedge fund that was massively down over its first 20 months. That's just not supposed to happen in the hedge fund world. Besides us, you can count the hedge funds that opened out of the gate down significantly for an extended period of time and survived and thrived using no fingers and no toes (if I've missed a major example I apologize). It just might be our proudest achievement.

Moonlighting

Another thing that makes our firm somewhat different has been our long-standing commitment to publishing and contributing to the general public discourse on investing. Over the years we have written many papers on quantitative investing (some on my own and some with my partners). Their early character was directly about our models, but in later years (with "Bubble Logic" the crowning example), they have focused more on issues of policy and overall market valuation. Issues like whether hedge funds really hedge as much as they claim, whether the most popular models for valuing U.S. stocks are broken, whether old style dividends are more important than people think, whether options should be expensed and whether international diversification gets a bum rap, have been some of our targets.

Speaking for myself, I have a tendency to tilt at windmills. In particular, when those supposedly "in the know" are pitching something (e.g., hedge funds, the overall stock market, or their own right to report mendacious earnings) based on illogical sound bites, falsely perceived expertise, or pseudo-intellectual flawed models, I admit I get a little crazy. I have been told that people with tempers generally have a problem in that they take too many things personally (e.g., the car that cut you off was probably not specifically cutting *you* off). I am guilty of this in my financial publications (e.g., an arrogant feeling of "I can't believe they think they can say this crapola about the Fed Model on my watch"). It's not particularly conducive to a Zen-like calmness for me, but it has fueled my passion for writing.

The result of many years of hard work has led to a body of work that I believe has been very helpful for investors, while it has integrated some humor with the substance, and in many cases, has taken highly contrary stands (e.g., telling your hedge fund brethren they don't hedge as much as they say and the lag in their marking-to-market overstates results, leading to lower net alpha than it first appears). If this sounds like I'm bragging a bit here, guilty, this is something I truly feel good about.

So, I'm skipping a lot now, but basically the last six years have been quite good for our process and our business. As mentioned earlier, AQR today manages about $35 billion and has a long list of clients that have shown confidence in us, which leaves us awed and humbled. We run well over 20 strategies (we started at Goldman in 1995 with four and at AQR in 1998 with seven), most of which apply value and momentum to a variety of investment decisions. And, we hope, there are still a lot more out there.

Geeks of the World Unite

Okay, enough autobiography. The careful reader will note that so far I have not really literally described how I became a quant. I'm not sure when exactly that happened to me as it probably happens gradually along the way. I've literally described "how I got to where I am" in the context of being a quant, which is similar, but not the same. While I don't think I'll ever precisely answer the title question, I think now would be a good time to try and tie up some loose ends by asking the question "what do we mean by quant to begin with?" You need to know that to discover when you became one!

Well, clearly we don't mean that nonquants are simply making qualitative judgments, as you don't see a lot of books with the title *How I Became a Qual*. I am not sure what that would even mean. Maybe they'd meet with management, and using no numbers, just invest saying things like, "I like the cut of his jib." Okay, nobody does that. So, is everyone a quant? Is the only distinguishing characteristic perhaps a certain lack of social grace and knowledge of 1970s and 1980s videogames?[20] Geek meets actuary?[21] Well, some of that may be true, but I think there's tremendous content to the distinction between quantitative and nonquantitative investment management.

So what is the content? Essentially, while other differences exist, much of it boils down to diversification. A quantitative investment manager's job is to find things that work on average. Usually (unless we're very lucky) the edge is small, even microscopic, for any one investment decision.[22] As such we usually or always (depending on the manager) follow our models precisely. Sometimes that's seen as "slavish" by quant cynics, but the alternative is nonsense. When you're following a model that makes thousands of decisions, judgmentally overriding (this does not mean you can't override for other reasons, like a suspicion of bad data) any one or a handful of decisions is highly unlikely to matter, and overriding many is impossible. And to the extent it matters, we quants worry very much that we'll undo what our models are trying to do. For instance, one hypothesis for value's efficacy is that it makes uncomfortable decisions that are ultimately rewarded. So, if you override the decisions you are most uncomfortable with, you risk undoing some or all of the point. Nonquantitative managers (I'm really going to call them *quals* for short, but I do admit they use numbers), by contrast, tend to know a lot about comparatively few situations. Their edge should be depth, and their handicap is they better be right, as they are not making too many bets. In effect, we (quants and quals) are both running casinos. Quants should have lots of tables with small edges, and quals should have far fewer tables with comparatively bigger edges.

One of my favorite anecdotes illustrating this came from my Goldman Sachs days before I started the Quantitative Research Group. I was listening to a conversation between a quant and a qual stock picker. The qual asked the quant if he owned a certain stock that the qual just added at a big weight and was very excited about. The quant replied, quite reasonably, "I don't know." The qual looked at him like he was from Mars.

Here's a short laundry list discussing my views on some specific points often made pro and con about quantitative investment (with my obvious bias showing):

- *Quants are data miners.* Yes this is an obvious danger of quantitative management as big computers and big databases make it easy to abuse this power. But, I think it's often overlooked that quals can be huge data miners also. If data mining is simply defined as believing that patterns you have seen in the past will repeat, without theory or intuition or logic, then quals can be as much a prisoner of what they have seen work in their careers as quants, and sometimes without as much perspective. The bubble of 1999 to 2000 is a good example. Many quals "learned" that tech/growth was the place to be from a career length run. They were wrong. I would say a form of data mining was certainly partially responsible. I think it's fairer to say that data miners make both bad quants and bad quals (with some exception made for some quants who pursue very-high-frequency strategies where something like data mining may be more appropriate).

- *Quants use "black boxes."* Again, perhaps with the exception of some very high frequency, sometimes AI–based pattern matchers, I know of no quants who can't tell you, with a precision that dwarfs a qual's story, why they own or do not own (or are short) each of their positions. The box is about as translucent as it comes. As I already noted, quants may not know everything they own, but once they try, they can tell you precisely why they own it! I think *black box* is a Luddite slur that is rarely accurate or fair.
- *LTCM's blow up shows the limits of quantitative investing.* I don't want to pick on the company, as I have a lot of respect for many of the individuals involved with LTCM, but its positions were very similar to the vast majority of Wall Street trading desks, none of whom you'd label quants. They might have been motivated by some quantitative theories, but they were not (to my knowledge) quants in the sense we are using the word—that is, investors who directly follow specific rigorous trading/investing models.
- *Quants have huge advantages over quals in terms of product design and risk-control such as market neutrality and separating alpha from beta.* This one, not shockingly, I agree with. I have taken, along with my colleagues, a rather public stance that much of the hedge fund world runs betas that are too high, follows implicit momentum strategies in setting their betas that reek of portfolio insurance, and generally are often charging alpha fees for returns driven too much by beta. Strangely, while you'd think a qual could just short a futures contract to get market-neutral, it doesn't seem to happen. Truly attempting to be market-neutral, and to more generally separate alpha from beta, seems to be a quant-only game.
- *Quants are "driving with the rear-view mirror."* This common broadside is true, but it's basically just the data mining critique again. More importantly, we are all (quants and quals) driving with the rear-view mirror. Good ones of both stripes won't just do what's worked in the past without a story and intuitive explanation for why it should work going forward. When someone says the aforementioned quote there is an assumption that quals know what's coming up in the future. If so, then somehow the evidence of that has not shown up after many, many years of looking. Quals also drive through the rear-view mirror, and I'll add that occasionally theirs is quite rose colored.

To summarize, I think good quant investment managers (like quals, there can be bad ones) can really be thought of as financial economists who have codified their beliefs into a repeatable process. They are distinguished by diversification, sticking to their process with discipline, and the ability to engineer portfolio characteristics.

So that's it. I always have trouble with conclusions, especially in an essay like this that doesn't argue a specific main point to be conveniently repeated. I will use my final words to say that I am a huge believer in quantitative management past, present, and future, again acknowledge a fair amount of luck in my own path, and, finally thank the partners, professors, clients, and family who made up a fair amount of that "luck"!

Chapter 15

Stephen Kealhofer

Managing Partner, Diversified Credit Investments

When our daughter turned three years old, my wife decided that she wanted to return to school to study law. Her choices came down to NYU or Berkeley. At the time, my academic career in the finance group at Columbia wasn't going anywhere. After a depressing tour of apartments and preschools on the Upper West Side, it was easy to choose Berkeley. In July 1985, we loaded up our little green Mazda with our possessions and drove across the country.

I had a visiting position in the finance group at Berkeley and was assigned to teach corporate finance. I went in to see David Downes, who was then the head of the MBA program. He was a gregarious ex-professor, tall and permanently freckled by the California sun, overwrought and overworked by the demands of students and faculty. I asked him what I should teach in the class.

"Teach a case class. Students will love it."

"But David, I don't even know what a 'case' class is."

"No sweat. Just get some cases. The students read them; you ask questions; that's all. They'll love it." He came around from behind his desk and patted me reassuringly on the back.

This seemed a little too easy. I cribbed some cases from a previous class and read them. I couldn't figure them out myself. In fact, I wasn't even sure if I got the drift of some of them. I couldn't see how anyone was going to benefit from reading them and trying to answer my questions. Instead I assigned them to groups as projects. I gave each one to two groups; one group to defend management's decisions and the other to critique them. I instructed the groups to use the actual case as just the starting point for their research and to dig up as much material as they could on the actual facts surrounding the particular corporate event.

This was my own entry point to practical corporate finance. Over time, I dropped the cases where I concluded I didn't know any useful theory or approaches. For instance, there was a case on dividend policy, which had originally been the basis for discussing various theories of dividends. Then one of my students called someone who had been on the board of the company and asked him why they had paid a dividend. The answer was that the company's founders were in a hole and needed some cash. So much for signaling theories.

The cases that I added were *Wall Street Journal* clippings of acquisitions, LBOs (leveraged buyouts), and spinoffs. The central issues were, increasingly, company valuation and capital structure. Capital structure was something that I likened to astronomers looking for Pluto. There was some mysterious force acting on Uranus; it must be another planet, so the astronomers went looking and found Pluto. In the case of firms, there was some mysterious force keeping them from being completely debt financed. We called it *the cost of financial distress*. Unfortunately, it had not yet been sighted. We knew, though, that it had to be intimately related to the probability of default. But no one seemed to really know very much about the probability of default either.

A Startup

Sometime in the fall of 1986, my brother's ex-girlfriend met me for lunch and told me about the company for which she was working. It was trying to create a vehicle for pooling bank commercial loans, much like mortgages were being pooled. I wasn't sure why anyone would want to do that, but she encouraged me to go talk to the principals, as they were looking for a researcher to do some consulting work.

The business was scattered through a small suite in one of the Embarcadero Center office buildings. I met with Michael Kieschnick, a young PhD about my own age, but infinitely more confident and experienced in the ways of the world, having previously been an economic advisor to California Governor Jerry Brown. He took me across the hall to a small, isolated office to speak with Oldrich Vasicek. I was in awe of Oldrich because of the groundbreaking work he had done on the term structure of interest rates. At that time, he was also working with Gifford Fong to bring the practical insights of that work to Wall Street.

Oldrich was a very handsome gentleman. Tall, lean, well muscled, sun-bronzed, casually but carefully dressed, at first you might mistake him for a professional wind-surfer. That is, until you heard him speak. He spoke in a high tenor voice with a pleasing middle-European accent, and his style was old-world, courteous, and charming. Polite, unassuming, reserved, he spoke carefully and in detail. One had the impression he had grown accustomed to explaining his work to those less intelligent than himself.

I am not sure how much I understood the first day, but the project that emerged was to test and improve a statistical rating model that Oldrich had developed. Surprisingly, it had predicted a default that Oldrich's implementation of the Merton model had missed. I had had little empirical experience or training, but I was confident that I could improve Oldrich's little model. Particularly as Oldrich had confided that he had developed it by reading all the papers in the area, collecting all the variables that they had used, and put them into an automatic program to find the best-fit model.

The project was my first real introduction to statistical data analysis. I was being paid by the hour. In fact, I was being paid more than I had ever been paid before, and threw myself into the project with enthusiasm, if not much efficiency. Two months later, I presented my results. I had isolated the variables that seemed to matter; I had considered a variety of specifications and transformations; I had developed several alternative models. And I had concluded that none of them were actually superior to Oldrich's little model. It was not the first time that I experienced his ability to get to the crux of a topic in what seemed in hindsight a simple and straightforward way.

Practical Defaults

It was several years later that a third window on the practical world of finance opened with a lunch with Pete Kyle. I had met Pete during a year I spent at the University of Chicago as a graduate student. He was a star there, and came to Princeton as an assistant professor while I was still trying to finish my dissertation. He graciously sat on my committee, and then by chance we both ended up at Berkeley at the same time. Pete was an expert in the information mechanics of markets, and he had agreed to do some work with a new litigation consulting business. They were looking for finance researchers. In short order, I was introduced to Tom Jorde, one of the two founders.

Tom was a professor at Boalt Hall, the Berkeley law school, and a former U.S. Supreme Court clerk. He was bright, outspoken, cheerful, and aggressive; ready to take on all challenges and win. What followed was a string of cases, mostly related to junk bonds: the U.S. government's case against Michael Milken et al., the bankruptcy of First Capital Holdings, and various S&L collapses. I analyzed data, reviewed business practices, and explained to the lawyers what the players were doing, and why. For me it was like instant experience. As I sifted through the endless memos and reports, I could follow the internal discussions in the subject businesses, review the decisions, analyze the motives, and evaluate the actions.

The downside was an enormous amount of reading and data to analyze. I remember a FedEx truck pulling up to my house and emptying its entire contents into my garage. For the case involving Drexel, we used the very latest PC technology available at the time—optimal magnetic disks—to manage the gigabytes of trading data that we were analyzing.

The commercial loan-pooling venture had not succeeded. But in the process of working on it, four important things transpired. First, when I tried to review the performance of the default prediction model that Oldrich had developed, it was clear that the limiting factor was the availability of data on defaults. At that time, no one kept data on defaults. Commercial databases generally purged firms that had defaulted. On the office wall we had a "Farside" cartoon that showed a frog sitting at a lunch counter and next to a diner who had just discovered something in his soup. The frog is asking, "You're not just going to throw that fly away, are you?" We were the frogs, and we went after the flies every way we could think of. For a while, we even experimented with calling firms to verify if they had defaulted. This did not turn out to be very useful, but it certainly led to some interesting phone calls!

Different groups of firms had very different default experiences. Once we began collecting default data, it became clear, if we were not going to be subject to arbitrary biases, that we needed to understand the totality of the default experience for a well-defined population. We set ourselves the goal of compiling a history of all defaults of all publicly traded companies in the United States, starting in 1978.

The resulting database revealed the second key finding. Despite the bad reputation that the Merton approach had at that time for pricing default risk, Oldrich's version worked quite well in predicting default. Further, the Merton model, because it was a complete cause-and-effect model of default, enabled us to understand how other approaches were working, and why they worked. In simple terms, the Merton approach reduced variables into three categories: information about the value of the business, the liabilities of the business, and the volatility of the business. Variables that were linked to default could be checked against their ability to predict those three quantities. Those insights enabled us to improve our implementation of the Merton model.

Finally, it dawned on me one day that our default database had become large enough that we could actually estimate empirical default probabilities. When we did that, we discovered that one reason the Merton model was not doing a good job of pricing credit was that the empirical probabilities did not match the theoretical probabilities coming from the model.

This process of work revealed something that has remained an ongoing theme for me in empirical work. The real world is a complex place, imperfectly revealed by noisy and confusing data. Progress was made by taking a position and using it to create an ordered view of the data. The view thus formed could never be independent of the initial position. Nonetheless, it could be useful and insightful. What I saw in contrast, in the work of others, was that people expected data to reveal answers even though they were unwilling to commit themselves to a position.

The third key finding was not an analytic one. Rather, having the opportunity to interact with bankers and discuss their process of credit origination and management, I could see the flaws in their process. Sandy Rose, who had been around banking for much longer than I had, had already seen the same issues and articulately put his finger on them in an influential series of columns that he wrote for the *American Banker*. It began with the fact that bankers did not understand the value of quantitative models. They were used to analyzing firms one at a time, looking for violations of norms. This whole mindset resisted the concept of generalization embodied in a model.

A banker formed an opinion, but could not quantify the process that led to that opinion. The result was that it was very difficult to measure the quality of the resulting opinions. Without that feedback, banks expanded their credit originations at their peril, because they could not ensure that default risk was being consistently measured at different places in their organizations.

Bankers rated risks; in other words, they ranked them. But they did not know the actual quantitative level of the risk. Thus, they could not know whether the price they were receiving for the risk was adequate for the level of risk. This led to a critical governance problem. A deal that could generate considerable accounting income was difficult to turn down on the basis of a subjective ranking. One banker described this to me as "hard drives out soft." It was hard to turn down $800,000 in revenue just because the credit officer did not like the credit. The credit officer's position became untenable once the competitors down the street were willing to do similar deals.

Finally, without a quantitative understanding of the risk, no real portfolio analysis was possible. Banks ran large portfolios; there was a great deal of diversification created within these portfolios, but no one could actually say how much was created or in what fashion.

The Entrepreneur

The fourth key event for me was getting to know Mac McQuown. Mac was a bigger-than-life character, who had a storied history in finance. In the mid-1960s, he had led an effort to rethink equity management at Wells Fargo bank. He had brought in Fischer Black and Myron Scholes, among many other finance notables, to work on the problem, and the ultimate outcome had been the creation of index funds. In that era he had met Oldrich, newly arrived from Czechoslovakia, and convinced him he should come work on finance problems at Wells (rather than analyzing dolphin communications, another offer Oldrich was contemplating at the time).

Mac had departed Wells in the early 1970s for the life of an entrepreneur. At the time I met him, he had been involved with starting a small stock index fund, Dimensional Fund Advisors, in Los Angeles, as well as a premium California wine producer, Chalone Group. Mac had a reputation as a bit of a wild man, but I was struck by his interest in people and the world. Strong as a bull, fond of fast cars and red wine, and newly remarried for the third time, he was then, and still is, a bon vivant of the first order. But he was also a great listener and a great talker, who enjoyed telling stories and spreading outrageous ideas. One of his great strengths was his ability to challenge people without being confrontational. He loved going into the proverbial lion's den and engaging, and I was often amazed at the quality of interaction that he stimulated.

I came to appreciate that he was also very thoughtful, always searching for data, rejecting conventional wisdoms and trying to get deeper understandings of the world. What might seem like a casual, controversial comment would turn out to be part of a closely reasoned, carefully constructed argument. He was surprisingly disciplined and energetic, able to pursue his multifarious (a favorite Mac word) interests and friendships while remaining highly focused on his objectives. He was the CEO of the loan-pooling venture, and as time progressed, I came to spend ever more time with him in conversation, sharing my findings and taking in his perspectives.

By early 1988 it was clear that the loan-pooling venture was not going to work. The venture capitalists were scrambling around trying to save something, in the process bringing in new management. I knew that we had something in the credit technology, but that it needed to be completely rethought and redone if it was going to be useful. In the spring of 1988, I went to Mac and Oldrich separately and asked them if they would be interested in starting a new business focused on creating a commercial credit risk service. My main argument was that we knew we had a general approach that worked, and further, that we knew that the banks and other credit-granting businesses desperately needed it.

What went unspoken was that we also knew that virtually no one in the market shared these perspectives. Few believed that an approach of this type could work, and virtually none understood how it could fit into their existing business practice. The exception was Lew Coleman, at that time the head of large corporate banking at the old Bank of America. He agreed to buy our first prototype product in 1989; without his vision and support we never would have gotten off the ground.

The business that we started in the fall of 1988 ultimately became KMV Corporation. For me, it brought together the disparate strands that I had been pursuing. From the litigation consulting with LECG, I learned about the operations of a variety of financial businesses, while getting a unique insight into the corporate bond market of the 1980s. From teaching, I had learned how to evaluate businesses and their financial decisions, and how to read their financial statements. From the work with the loan-pooling business, I had been propelled into large-scale data analysis, as well as into fascinating first-hand encounters with financial businesses that I had previously only known second- or third-hand.

Moreover, the subject matter had invariably centered on a key variable: the probability of default. We could make much more sense out of capital structure if we could know the default probabilities. We could value corporate debt if we knew the default probabilities. We could begin to quantify portfolio properties if we knew the default probabilities. From where I stood, I felt like I had just set foot on a lost continent.

Inventing a Business

Now the real work began. We set up shop in a small office carved out of what had been the Del Monte Milling Warehouse built in 1907. Shortly after we moved in, the owner filed for bankruptcy, and over the subsequent few years, we watched the other small businesses around us fail as the 1991 downturn took hold of the San Francisco economy, leaving us almost alone in the building. As a result, the rent was very good. So was the address: Montgomery Street, the center of the San Francisco Financial District. In this case, however, rather than being in that district, we were a mile north, tucked under the lee of Telegraph Hill on a funny little stretch of Montgomery that started under a cliff and ended two blocks later at the water.

In addition to Mac, Oldrich, and myself, by mid-1989 we had added Peter Overmire and Rob Rudy, former colleagues who were interested (and crazy) enough to join our little venture. Unlike the previous venture, we decided not to seek any outside financing, so money was very tight in the early days. The founders engaged in various outside activities to produce income, plus I brought most of my consulting work to KMV, which enabled us to pay some modest salaries.

We rethought and redeveloped our models, focusing on conceptual clarity and empirical verification. We needed names for our variables, and I remember sitting around brainstorming with Oldrich about what to call various variables. We coined *expected default frequency* (EDF), *loss given default* (LGD) and others on the principal that a three-letter abbreviation was the best identifier. It was always amusing years later to see these crop up in others' work.

We were one of the first businesses to use the new CRSP NASDAQ file, and spent months mapping our default history to COMPUSTAT and CRSP history. The technology infrastructure was primitive; our "network" consisted of cables looped over the top of overhead drainpipes, connected via LapLink. PC processing times were slow, most of our code written in Fortran, and our database consisted of large flat files. But it worked.

Our first "product" we nicknamed the *phone book*. It was a hard-copy listing of default probabilities for all five thousand publicly traded firms in the United States. Besides the default probability, it also had a little arrow next to the number, showing if the value had recently gone up or down. It was about as easy to look at as a phone book, and I can't remember if we managed to ever get anyone to pay for a copy.

A visit to some old acquaintances at the Bridge brokerage in St. Louis opened our eyes. They had wanted to show us their analytics, which we took to be something like the complex number-crunching we were doing. When we got there, we didn't see any sophisticated analysis, just graphs of prices and other data. It suddenly occurred to us that these were their *analytics.* Moreover, unlike our output that looked like, well, a telephone book, theirs consisted of beautiful pictures. I went back to San Francisco, and on a few sheets of yellow notepad pages sketched out a series of graphs to display. Within a few months Robert Rudy had figured out a software version and had produced a working prototype. Not long after that, we had our first real sales.

It became evident to us that it was difficult to sell our product to most credit officers, because it was an affront to their own traditional methods. At that time, Mac was doing most of the selling, and when he returned from a business trip, we would gather around and discuss the implications. One of his cautions was to not be hasty in concluding that our product was unsaleable, but rather take the time to understand the perspectives we were hearing. Ultimately, Mac hit upon the idea that we should focus our efforts around the topic of bank portfolio management, where the banks had little or no current activity, but were very interested in developing a portfolio management function.

This proved to be a very successful tactic. For years afterwards we would have individuals show up on our doorstep announcing that they had been appointed the bank's first portfolio manager, and wanting to know what they should do. The steps always included needing default probabilities. In this fashion we introduced our EDFs into the banks' credit decisions.

Portfolio Management of Credit Risk

In the beginning of bank portfolio management, we did not have a very good idea of what to do either. We had a theory, some difficult mathematics, but little connection to the empirical characteristics of actual bank loans. Sometime in 1989, Oldrich and I had discussed the feasibility of actually dealing with bank loans in a portfolio model, and concluded that it wasn't feasible. The difficulty was that banks did not really treat loans themselves as specific contractual claims, but rather, as obligations arising from an ongoing and often ill-defined relationship with its borrower. Moreover, they contained a variety of implicit and explicit embedded options that were not well specified or consistently exercised by either bank or borrower.

Despite our reservations about modeling bank portfolios, we convinced ourselves to tackle the topic. At the time, I suggested the submotto for our business should be "Fools rush in . . ." Again, Lew Coleman at Bank of America, as well as new supporters at the old Chemical Bank, came to our rescue by signing up for portfolio assessments well before we knew what data could actually be obtained on bank loans, or even what data we would need to effectively describe loans.

There were a variety of lessons in the exercise. First, it was not primarily an exercise in *financial engineering,* a term then coming into vogue. One could not divorce the modeling aspects of the problem from the nature of the available data, nor from the business context in which the model was to be used. Our overriding criterion became making sure that we produced business advice that would improve business practice. Mac and I spent much time trying to understand what banks were doing and why. It was a bit like anthropological research in that the proffered explanations did not match the practice, so observation was as critical as discussion.

Just like anthropological research, the study was often fascinating. One sometimes forgets that finance is social science; it is about the behavior of individuals in the context of organizations and markets. Also, that producing change in organizations requires consideration of the motives, information and incentives of individuals within those organizations. In marketing, we focused our attention on the education of individuals, rather than on some ill-defined larger unit of organization. This had payoffs for us in three ways. First and foremost, we got to know a fascinating cross-section of people, with interesting and unique perspectives, one of the primary sources of fun in the business. Second, clients or potential clients would move around from one organization to another, and serve as agents of propagation for the ideas we were developing. Third, as these ideas became established and proved themselves in practice, it was rewarding to see the individuals who had bet on them move up to positions of greater influence.

The second thing that I came to appreciate was the old adage, "The pen is mightier than the sword." I don't think I had really appreciated the power of ideas. In the context of business, there is a real reward to pursing better ideas. By being able to translate abstraction into concrete business practice, we were able to unleash the power of new ideas about credit and portfolio practice. It made me realize that virtually any obstacle can be overcome if there is, in fact, a better way of doing business. The primary challenge is in defining practical steps that a business can take that move it from its existing practices to new practices while not incurring excessive cost or risk along the way.

The third principle was an elaboration of the idea that you have to take a position in order to generate a view. For instance, banks were often engaged in trying to measure the profitability of some area of their businesses. They treated this as an exercise in cost accounting, where they would assign costs and revenues to the particular activity, and then conclude that it was profitable or unprofitable. I came to realize that they could not know the profitability by this measure, because the very lack of knowledge that generated the measurement project meant that the area of business was not being managed to maximize profitability in the first place. The business problem was not to measure the profitability but to figure out how to do the business profitably. Only then could they conclude that it was profitable or unprofitable.

This principle has many corollaries. You cannot know the value of a business without having a view on how to utilize its assets. I think of this point whenever someone comes running to me with the claim that a company is over- or undervalued. The right question is, whose view of the assets is being valued?

In our case, banks did not have a portfolio management practice. You could not define an optimal practice in the abstract because there were too many unanswered, and unanswerable, questions about the characteristics of loans. Nonetheless, by taking a position on the answers to these questions, it enabled a process to take shape that would begin generating data by which one might be able to answer some of the questions. Otherwise, no progress could be made.

An example illustrates this point. When we first began working with the old Chemical Bank, one issue was, is the credit quality of a loan to a subsidiary enhanced because the parent has expressed some intention of supporting the loan? This support was expressed in the form of what bankers call *comfort* letters. It turned out that in Chemical's database, it had 28 different categories of comfort letters. We looked at a variety of complex ways of modeling such support. Fortunately, with some good external guidance, we took the view that there were not 28 categories of support, but rather two: 100 percent support or 0 percent support.

This position invariably caused heartburn with bankers, because they were used to ducking the issue. But the truth was, they had no data to support an alternative position. The comfort letter was just that, a way of creating "comfort" for someone so a deal could be approved, but no one had conscientiously examined how support from a parent in the event of default actually played out.

It was not that our position was correct. But by taking it, we forced the bank to define its processes with respect to this question. In this case, the result was a process that specified what effect a particular parent-subsidiary relationship would have on the default probability and loss-given-default assumed for the subsidiary loan. This, in turn, affected the risk of the loan in the portfolio and thus its perceived profitability.

There were two possible outcomes that flowed from this specification. The lending officer could come back and make a more specific case why the profitability assessment did not make sense, or with the passage of time, data could accrue that would support or undermine the position taken. In either case, one would subsequently modify the process to account for the new or additional data. In the absence of a defined process, you could not generate the data needed to make the assessment. At the limits of our knowledge, we have to take a position in order to create a view.

A Room with a View

By 1993, our business was expanding and we had moved upstairs in the warehouse into a larger space. Its only drawback was that the previous tenants had used the conference room to house the litter box for their pet cats. Whenever we had visiting clients, we were always hoping they didn't notice the lingering taint.

By this time, we had a working portfolio model for bank credit risk. As we began selling it, it raised many questions around the subject of bank management in general. The result was that our discussions at banks ranged from low-level data integration to finance theory to high-level discussion of business strategy.

One amusing example of the type of interaction was provided by a group from Goldman-Sachs that came to our office, with Fischer Black on the phone in New York. In the middle of discussing the portfolio model, Fischer interjected, "Banks don't need to diversify their portfolios; their portfolio risks can be diversified by equity investors."

This is, of course, a fundamental proposition in finance, and, while it has an answer on several different levels, involves a complex discussion. There was silence around the table. The Goldman bankers were clearly not interested in getting into that discussion. Their eyes came to rest on Bob Litterman, the one who worked most closely with Fischer.

"Well, Fischer, that's right. But the banks *want* to diversify."

"Oh, well, if they want to diversify, then that's Okay." The discussion resumed with looks of relief around the table.

We had not been sure when we started the business whether we were selling models, or data, or a service, or some combination. In time it became clear to us that we were primarily valued as change agents. The models and data were critical to being able to make the possibility of change concrete, and to actually effect it in practice, but the real value lay in assisting banks, and ultimately other financial businesses, change their business direction. We developed a business model that supported this relationship, and that was key to our success. The package that a business sells, by which it adds value for others, is a more complex proposition than a simple product.

We had also begun to work in Europe. A recently formed consulting business, Oliver-Wyman, adopted much of our technology and introduced us first to U.K. banks, and then elsewhere on the continent. This forced us to redevelop our models for these markets, based on the data then just becoming available. It also led us to develop a broader perspective on the institutional development of banks and markets around the world. Ultimately, it afforded us an unparalleled view of global financial integration. By the time we sold KMV to Moody's in 2002, we were tracking, on a daily basis, upwards of 30,000 firms in 40 countries.

As the business grew, we added employees. In explaining ourselves to new colleagues, we used to joke, paraphrasing an old Navy recruiting poster, that our objective at KMV was to "have fun, make money, change the world—not necessarily in that order." In time we came to realize that it was perhaps the best description of the business.

Finance is a beautiful field. First, it touches everything, so it is a license to explore the world. Second, there is more data in finance than any field I can think of. One of the many great things about markets is that they generate large amounts of high-quality data about the actions of people. Third, the path from data analysis to the implementation of business ideas is incredibly fast. In biotechnology, it is measured in decades; in finance we do it in months or even weeks. Fourth, if you can deduce something useful, you get paid.

Twenty-five years ago, I was fresh out of economics graduate school, interviewing at a truly odd collection of schools, when Charlie Wolf swooped in and made me an offer to come to Columbia University and do finance. Wherever you are Charlie, thank you.

Chapter 16

Julian Shaw

Head Risk Management & Quantitative Research, Permal Group

I was getting desperate at the University of Toronto in 1986, where I was working on my (still incomplete!) PhD thesis in computational complexity. Godel's theorem established *absolute* limits on what we can know: Any formal system powerful enough to express arithmetic contains either contradictions or propositions for which there is no proof in the formal system. Computational complexity is concerned with *practical* limits on what we can know. For example, if there is no algorithm that can solve a generic problem in time proportional to a polynomial function of the number of variables in instances of the problem, then in general we will only ever be able to know the answers to small problems of that type.

My advisor and others established that for a huge variety of generic problems, there are algorithms that check proposed solutions in time, which is a polynomial function of the number of variables. Yet the best-known algorithms for actually *finding* solutions to such problems run in time, which is exponential in the number of variables. Furthermore, if any of the generic problems have a polynomial time algorithm for finding solutions, then they all do: Hence, Cook's famous conjecture that there is no such algorithm, i.e., $P \neq NP$.[1]

Unfortunately, the only problems I found interesting were too hard, or at least too hard for me (a good sign that one is in the wrong field). I wasn't getting anywhere, and my money had run out. As a foreign (Australian) student in Canada, I wasn't permitted to work outside the university. So when the finance department at the University of Toronto advertised for a computer programmer on the intranet, I jumped at it despite my ignorance of finance. They gave me a copy of Levy and Sarnat's *Portfolio Theory* and told me to write them a mean-variance optimization program.

Mean-variance optimization is one of the twin glories of quantitative finance (the other being the Black-Scholes option pricing model). Unfortunately, it doesn't work well in practice. The users of my program complained that the "optimal" portfolios produced by my program were grotesquely underdiversified. In fact, the program correctly produced optimal portfolios *given the parameter values* (the means and the covariances of the assets). Unfortunately, all we have are estimates of these parameters subject to statistical error, *about which the optimizer knows nothing*. Despite the efforts of some (e.g., Jobson and Korkie, Jorion, Michaud, and Scherer), there still isn't a completely satisfactory solution to the problem of incorporating this estimation risk into the problem. It's a problem I wrestle with today when creating optimal portfolios of hedge funds.

My mentor at the University of Toronto was Myron Gordon, professor of finance and Fisher Black's immediate predecessor as president of the American Finance Association. When the optimization project finished, Gordon gave me work assisting his research into the cost of capital for utilities. Gordon often appeared at public hearings on utility rates, arguing on behalf of consumers against tariff increases. Not the side finance professors usually take!

Gordon is the developer of the Gordon Growth Model; but one of his less conventional contributions to finance is a model of the development of capitalist societies via simulated investment strategies. In his model, Gordon assumed that the only available investments are a stock portfolio with parameters similar to that of the S&P 500, and a bank account. People are free to adjust the portfolio mix, but their consumption exceeds the risk-free rate so they must invest some fraction of their wealth in the risky asset in order to survive in the long run. Those whose wealth falls below a threshold have to work for a living.

From this work emerged a new investment criterion, maximization of the probability of long-run survival, which is qualitatively different from the other famous normative[2] investment criterion, the Kelly (log optimal) criterion. Some conclusions: The class composition of society can be explained by dumb investment luck, rather than the skill or virtue of the capitalists, and closed capitalist societies are unstable in the long run.[3] Another of Gordon's projects was teaching finance in China as a practical contribution to the well being of the world's people.

From this distinguished but unorthodox finance professor, I learned that finance is more than just pricing models and arbitrage, more than picking up dimes in front of steamrollers and sweeping up golden crumbs.[4] Finance is concerned with the way people organize production and distribution, which is a prerequisite for human welfare. It is not inherently reactionary. Although the principal use of quantitative finance is making money out of market inefficiencies, it can also be used to explain and guide economic behavior. Albeit it rarely is.

For example, recent changes to accounting rules force firms to include accurate valuation of their pension liabilities in their balance sheet. Consequently, firms are replacing their defined benefit schemes with defined contribution schemes under which each worker acts as his own portfolio manager. Personal portfolios tend to be much riskier and less diverse than optimal and are burdened with high transaction costs. The long-term result of amateur portfolio management, combined with increasing longevity, can be modeled: old-age poverty for millions of retired workers. But this is not the kind of research most quants get paid for.

Gordon Capital

I got married in 1987, and I needed a full-time job. I could have found a job teaching math or computer science at a cornfield university while finishing my PhD But by now I was hooked on finance and big cities, so I was drawn to Bay Street, Canada's equivalent of Wall Street. I eventually found a job as the sole quant in a firm of aggressive market makers in Canadian equities, options, and futures, Gordon Capital (no relation to Myron Gordon). Gordon Capital was famous for its willingness to take risk and pioneered the bought deal in Canada.

My acceptance at Gordon Capital was grudging. "Conehead" was my nickname, and I remember the pugnacious founder of the firm asking me for my proposed hedge for an impending equity portfolio trade. When I started explaining beta adjustment and tracking error, he stopped me. "Here's my hedge," he said, and with a flourish he pulled out a piece of paper with a handwritten list of the stocks in the portfolio and the prices institutions had pre-agreed to pay for them. I had to agree his hedge was better than mine.

Even the options traders were skeptical: "You quants come along with fancy new pricing models, but your model prices are inside the bid-ask spread! How are we supposed to make money from that?" They had a point; a lot of fancy option pricing models are like this.

Gordon Capital conceded it needed a quant for structured derivative valuation, which was fortunate since Gordon Capital pioneered structured products in Canada. American retail investors wanted to bet that the Nikkei would fall from its then-stratospheric heights, but there was nothing they could trade (the U.S. government, which upholds the right of its citizens to bear arms, draws the line at *derivatives*). The loophole was that American investors could trade Canadian warrants after the warrants had traded for 90 days in Canada, so Gordon Capital created Nikkei put warrants, quantoed into U.S. or Canadian dollars.

I remember smugly marveling at the huge implied volatilities those dumb investors paid for the warrants. However, the dummies got the last laugh when the Nikkei finally crashed. (Fortunately, it was not at our expense. We were hedged.) This taught me that there a lot of different ways to make (and lose) money in finance.

Things ended badly at Gordon Capital due to deals it was my privilege to unravel. The essence of these deals (as is the case with so many supposed arbitrages) was the sale of insurance. Gordon Capital didn't realize it at the time, but it was short billions of dollars of Government of Canada bonds and long a portfolio of high-quality bank paper that had been carefully constructed by an advisor who used a linear program to match the coupon flows. (The advisor has since been indicted on unrelated charges.)

To use an Australian expression, this was "a nice little earner" until bank credit spreads deteriorated and Gordon Capital wanted to reduce its exposure. To the firm's surprise, it could not do so without sustaining a huge mark-to-market loss, and the whole scheme was blown open. How could this happen? First, the documentation was very complicated, and some of the documentation had the effect of negating clauses in other documentation. One had to read the full set of documents to figure out what was going on, and even then it was hard going. Second, and I have seen this repeatedly in my career, while a business appears to be making money there is little appetite for analysis of the apparently golden goose. Firms that blow up often have smart people working for them. It usually turns out that they weren't asked to look at problems until too late. Enron and its once highly touted risk-management department springs to mind.

Since then, I have spent much of my career extracting the financial essence of complex legal documentation. Here's a tip from my father who was a solicitor before he became a physician: nonlegal people often think that legal documents do not really mean what they say. This is a fallacy. Legal documents usually do mean what they say, and anyone who ignores what a legal document says is a fool.

What does this have to do with quantitative analysis? Usually the hardest part of quantitative analysis is framing the problem in the first place. Many quants aren't good at this, partly because they don't think it's their job. They want the PDE or whatever to be served up to them on a plate. Solving PDEs is a useful skill, but there is a lot of good software to do that. There is no good software for identifying options embedded in deals defined by verbose legal documents. Derivatives training gives one a huge advantage; you can find options everywhere *if* you look hard enough. For example, the essence of the deals I have just described was a strategy of writing out-of-the-money options. (Some successful hedge fund managers run essentially similar strategies!)

CIBC

My next job was an unhappy sojourn at CIBC Wood Gundy, as vice president for Interest Rate Derivatives. The lowlight of my stay there was a dispute between CIBC and a trader over the valuation of his book. The dispute was referred to the unlamented auditing firm Arthur Anderson (put out of business by the Enron fiasco), which independently marked the trader's book of caps, floors, and swaptions *using a flat volatility surface.* The trader got a million dollar check out of it. The moral is that independence without expertise is worse than useless.

Things haven't changed much, people still think that having complex instruments marked by an untrained, overworked clerk in Curaçao who gets his prices from a broker who earns his living from commissions paid by the creator of the instrument is reassuring.

Barclays Capital

One day, at the end of 1994, a headhunter called from London and tried to interest me in a risk-management job at Barclays Capital. I wasn't interested. Gordon Capital had no risk management and CIBC (pre-Bob Mark) had ineffectual risk management. (If you don't know even know how the books are marked, how can you know what the risk is?)

The headhunter persisted; this was a position in a new-style risk-management department, and the boss of it didn't want old-style risk managers. Skeptically, I agreed to a lunch interview in New York, at which I met and had an argument with Colin Lawrence. At the time, all trading desks had the same attitude toward risk management as I had at CIBC. To break through this and establish effective independent risk management required strength of character, and Colin, a proverbial bull in a china shop, has it. A few days later, to my surprise, I was offered the job as head of derivatives risk at Barclays Capital in London. Intrigued by Colin, I took it. Colin was a great risk-management pioneer, and many people owe their first risk-management job to him.

On the whole Barclays was fun, but establishing a risk-management function was a challenge. Barclays had smarter people than I was used to dealing with; amongst them Riccardo Rebonato, Mike Sherring and Paul Doust. In any case, I was now on the other side of the fence; I was now one of those dopey risk managers I had learned to despise.

Fat Tails and Thin Peaks

One of the quant traders was much better at algebraic manipulation than me. Paradoxically, this was to be the source of his downfall and provide a moment of glory for me. This trader loved closed-form solutions: in his mind, numerical solutions were for sissies. His specialty was interest rate derivatives, and although he had never traded FX options, he derived closed-form solutions for single and double barrier FX options and began trading them in huge volumes. I was convinced that he was paying too much for them. When I discovered that the counterparty for most of his trades was a subsidiary of Goldman Sachs (such was the trader's faith in his own models he neither knew nor cared), I put my foot down and got trading halted.

The problem was that in order to derive closed-form solutions, one generally has to work in the Black-Scholes framework. Everyone knows that, due to the fatter than lognormal tails in most asset returns, far-from-the-money options should generally be priced significantly above the Black-Scholes formula price. But these barrier and double barrier options were close to the money, so this problem doesn't apply, right? Wrong; if a distribution has fat tails then *it must have a taller thinner peak to compensate*. Thus a formula derived in the Black-Scholes framework must price near-the-money barrier options too expensively. This is the crucial intuition that the trader had missed; departures from log-normality are important for pricing barrier options, but he had left them out because of his devotion to closed-form formulae and the assumptions that they entail.

No risk manager had ever stopped trading at Barclays, and the FX option salespeople wanted to kill me. The normally aggressive Colin Lawrence looked worried when I told him what I had done while he was on vacation, and after a week trading resumed. At least I slowed them down, and a few months later I was vindicated—the book blew up. But the firm lost money, so this episode must count as a risk-management failure.

Risk management, at least in investment banks, has moved on. Good risk management works with businesses to create the infrastructure needed to support new products. Arguments about pricing, hedging, and risking should take place *before* trading begins, not after.

Adventures in CDO Land

Another incident shows the power and limitations of quantitative risk management. A charismatic ex-derivatives trader started a CDO (collateralized debt obligation) business, which he ran as a derivatives business. The accounting was mark-to-market, and he acted as both the structurer and the active manager of the portfolios of the underlying portfolios of credits. A general problem for CDO structurers is that they have to find buyers for all the tranches of a new structure at once in order to launch the deal. My CDO trader neatly solved this problem; he simply referenced unsold tranches of new deals into his previously created CDOs!

Although one could quarrel with some of the inputs, his models were not bad, or at least no worse than the industry standard Normal copula model. But the business model was bad: acting as both the structurer and the portfolio manager left the trader hopelessly conflicted. There was no way to prove that his credit substitutions were for the benefit of the tranche holders. Thus, in effect it granted the buyers of tranches a free option to get their money back if their tranche didn't perform—which they exercised.

A fascinating aspect of the financial world is that people with diverse approaches can coexist in the same market. For example, my CDO trader was enthusiastic about quantitative models, but there are nonquants in the CDO structuring business who scorn them. Such managers usually own a portion of the equity tranche personally, supposedly to align their interests with those of the other tranche holders. They are surprised to learn that *increasing* the correlation between the reference assets (while maintaining the average credit quality) transfers value from the senior tranches to the equity tranche. So the supposed alignment of interests is approximate at best!

Nonquant managers resist marking to model: I once modeled a CDO equity tranche. My model said it was worth about 27 percent, but the nonquant manager had it on his books near par. (Subsequently, the tranche traded near my model price.) This nonquant manager is still in business, in a hedge fund whereas the "smarter" quant trader is not. This goes to show that sometimes the business model is more important than the quantitative model.

The Strange Evolution of Value at Risk

In the beginning, VaR calculations were usually based on the assumption of a multivariate normal distribution of a large number of market risk factors, a method generally known as *variance-covariance* or VCV. The variances and covariances of these factors were estimated with complex GARCH models (or particularizations of GARCH such as exponential smoothing).

What is the situation today? Are today's VaR models even more complex? No, they are much simpler! Almost everyone, even at JP Morgan where the multivariate normal approach was invented, has switched to an approach so simple even my mother can understand—historical simulation, or *HistSim*. Why?

First, for statistical reasons estimating big covariance matrices is hard even if the market is normal. Second, the market isn't normal, so mean variance VaR breaks down around the 95th percentile—before things get interesting.

But there is a subtle and more important reason for the triumph of HistSim over VCV despite HistSim's crudity: *mean-variance VaR is hard to audit but historical simulation is easy to audit.* By "audit" I don't mean what audit departments do. I mean explaining to an irate and skeptical trader exactly why you think his or her portfolio is as risky as you say it is.

There will always be mistakes in collecting data, so a key requirement of a risk system is the ability to drill down and diagnose the results. If one says to a trader, "If you had the position you have today then on each of these specific 10 days in the last two years you would have lost more that $10 million," then he or she can understand this and contest it. The great virtue of historical simulation is that it is easy to *resolve* arguments. Either the trader is forced to agree he really would have lost the specified amounts on the specified days, or you are forced to concede that either the position data or the historical market data are wrong.

By contrast, if you say, "I have applied a fancy VCV (or even fancier) risk technique to your position and this is the VaR number that emerges," then resolving the disagreement is infeasible. The result: ineffectual risk management. *A crude but transparent risk system is more effective in practice than a sophisticated but opaque system!*

A Paradox

What do you call a quant who works on really important financial problems? A *business*person. Strangely, as a rule the quantitative firepower applied to problems in finance is inversely proportional to the amount of money at stake. For example, exotic derivatives trading desks get a lot of quantitative attention, but the risk exposures of exotic derivatives desks are dwarfed by the credit exposures at the same banks, which are only now receiving quant attention thanks to the requirements of Basel CAD (Capital Adequacy Directive) II. In turn, these credit exposures are dwarfed by those of pension fund portfolios, which are run by amateur trustees with advice from actuaries who know little about finance (hence, the pensions crisis).

Quants congregate in derivative valuation, yet there are other areas of finance that are crying out for decent models. Yes, the problems seem "fuzzy." The first reason for this is we haven't solved them yet! Before Black, Scholes, and Merton showed people how to do it, option valuation was a fuzzy problem, too. The second reason is that financial data are sparse and noisy. Quants need to retool to address these problems. For example, I saw a trading desk tell the quant department it did *not* want yet another exotic option model, it wanted better volatility forecasting techniques for the existing models. The trading desk didn't get them. Unfortunately, many quants from the physical sciences are too lazy or arrogant to learn some econometrics.

Permal

Risk management at an investment bank was more fun when we were creating the role. Hence, three years ago, when I was offered my current job as the first dedicated head of risk management at Permal, one of the largest funds of hedge funds, I was very excited. Since no one really knew what this entails, I would have to make it up as I went along, as I had done 10 years earlier in an investment bank.

And that is how it has turned out. Figuring out the risks that diverse hedge funds take to make their returns, figuring our which hedge funds actually have skill (most don't), combining the good ones into portfolios and developing new risk management techniques for hedge funds is a fascinating job.

What Makes a Good Quant?

Identifying and analyzing the key components of business problems and choosing the right techniques to model them are the hardest parts of quantitative financial analysis; the actual application of the quant techniques to problems is usually not very hard, or at least, it usually doesn't *need* to be very hard.

First-rate minds make significant discoveries in mathematics and science. People become finance quants after they realize they are unlikely do so. Finance provides gainful employment for dilettantes with nimble second-rate minds. But to become good tradesmen and tradeswomen, quants must *unlearn* two attitudes that one absorbs in academia.

The first attitude is that specialization is good. Bad quants pretend they are still physicists or mathematicians or whatever; they can't be bothered learning new tricks from other disciplines, and they never learn much finance. "If all you've got is a hammer everything looks like a nail."

The second attitude to unlearn is the assumption that the failure of people to understand what you are saying proves how smart you are. In academia, one only cares what the other specialists in one's own field think. By contrast, in a financial institution nothing you do counts for anything unless you can convince the (usually) intelligent but nonmathematical people in charge to do something different from what they would have done otherwise. Despite Poincare's famous maxim that one doesn't really understand a mathematical theorem until one can explain it to the first person one meets on the street, most mathematicians can't even explain their theorems to other mathematicians! And they are proud of it!

There is a discipline to science and mathematics that is different from other human endeavors. When you write a scientific paper, you have to prove what you say. If there is a hole in your proof or a flaw in your experiment, your competitors will shoot you down. Cold fusion and stem cell cloning didn't survive long.

But quantitative finance is also subject to a discipline—the discipline of *making money*. If your idea is wrong or not implemented properly, a lot of money, not to mention your job, may be lost. The old jibe "if you're so smart, why aren't you rich?" is appropriate in a financial institution. Some quants overcome dysfunctional attitudes learned in academia, play important roles in their businesses, and become rich. Others relegate themselves to the back room forever.

The Art of Leaving Things Out

Quantitative finance is a craft and a trade. It is barely engineering, and it is certainly not a science. Good quantitative finance can be summed up as the art of leaving things out, plus the art of selecting the right tools.

One can't model everything. A good quant lets the problem tell him or her which features are economically important and must be modeled and what can be reasonably left out or approximated. For example, as we saw, to construct optimal portfolios one should address parameter estimation uncertainty; it can't be left out. For another example, in pricing barrier options, departures from log normality are important and should be modeled.

The opposite error is to throw too much in. This accounts for some of the baroque interest rate derivative models with myriad parameters for which there is no good estimation procedure. As Riccardo Rebonato says, the test of a model is its *ability to hedge* a real portfolio in the markets (and this includes parameter estimation), but quants often don't listen. I remember a trading desk thinking it had done most of the work by creating a fancy jump diffusion model with myriad parameters. The desk left estimation of the parameters to the risk management department. They should have realized that parameter estimation is *a critical part of the problem*. And they should have checked their model's ability to hedge a portfolio using historical data before they used it in real time with real money.

The Art of Choosing the Right Tools

One can't be an expert at everything but one does need to understand the essence of a wide range of mathematical and statistical techniques. It is rare that one needs to invent completely new techniques; usually you can find a problem with the same mathematical form, often in an apparently unrelated discipline. The art of the quant is to find appropriate tools and bolt them together to create an effective solution to the problem at hand.

Virtually every mathematical technique has basked in 15 minutes of financial quant fame: stochastic calculus, numerical PDE techniques, wavelets, neural networks, extreme value theory, GARCH, chaos, and so on. But the master of any one technique is often a poor quant. One still finds people and books that essentially say that finance is a just a trivial application of a single technique—the technique in which they happen to be an expert.

One can usually get most of the benefit of specialized techniques from a fraction of the theory. But experts in each specialization shamelessly exaggerate the benefits of improvements and refinements to the theory and, therefore, the need for their expertise. Many of these improvements and refinements don't work or aren't worth the effort—I've tried a lot of them.

Extreme value theory is a good example: For practical applications of EVT in finance all one needs to understand are two simple principles: (1) Model the tail directly rather than fitting a distribution to the entire data sample and hoping the tail comes out OK; (2) model the tail by simply fitting a straight line to a log-log chart of the tail of the empirical cumulative distribution function. That's all there is to it. There is no need to get bogged down with fancy estimators—they all perform about as well as using a ruler and log-log graph paper or the computer-programmed equivalent.

More examples of techniques that have easy and effective financial applications:

1. *The Kalman Filter.* Developed in the 1950s by engineers for tracking spacecraft, this algorithm is very useful for tracking sensitivities to financial risk factors. It can be regarded as a time-varying generalization of regression and works much better than the moving windows one finds in many finance papers.
2. *Resampling Statistics.* Real scientists never estimate anything without simultaneously computing *confidence bounds*. Much stupidity in quantitative finance is due to failure to specify the uncertainty associated with estimates of financial quantities. With modern fast computers, there is no excuse for this; it is simple to create error bounds by

resampling from the original data with replacement. If quants only knew how fuzzy many of their inputs are, they wouldn't waste so much time on ultra-refined high-precision models. But then again, perhaps they would, because ultra-refined high-precision models are what they are good at. If all you've got is a hammer . . .

3. *Copulas.* Copulas provide an example of the haphazard evolution of quantitative finance. The key result is Sklar's theorem, which says that one can characterize any multivariate probability distribution by its copula (which specifies the correlation structure) and its marginal distributions (the conditional one dimensional distributions). Thus one can create multivariate distributions by mixing and matching copulas and marginal distributions. As far as I know, copulas were introduced into finance in 1999 by David Li[5] at RiskMetrics purely as a computational device to speed up CDO tranche valuation, but the only copula used was the normal copula. Inadvertently, Li's paper opened people's eyes (or at least my eyes) to *non*normal copulas, which are very useful in financial risk management because one can create and fit multivariate distributions that have higher correlation in extremes, a well-known property of financial markets.[6] Why didn't we know about them before?

Do Quants Lack Business Sense?

Despite the now widespread employment of quants in finance, few have made it to the top. Fisher Black was a partner at Goldman Sachs, but he was the exception to the rule. Perhaps this is a good thing, since quants often lack business sense.

I can think of two spectacular examples. The first example is Enron's quantitative traders exploiting California's electricity trading rules using quantitative strategies with juvenile names such as *Death Star*. What they didn't understand is the unwritten rule that overrides all the other rules: If you screw enough people, eventually they will get you back. Have people learned since then? No, as the recent Citibank European government bond debacle, complete with a juvenile name for the strategy (*Dr. Evil*), demonstrates.

But the most spectacular example of all is Long Term Capital Management. Here we had the two most famous living quants supposedly helping to run an enormous hedge fund, which got blown to pieces when credit spreads blew out. Jorion provides a devastating analysis of LTCM's risk management mistakes.[7]

Incidentally, LTCM demonstrated an enduring belief, that there are lots of really smart people who could make tons of money if only they weren't inhibited by pesky regulators, unimaginative boards, and overbearing risk-management departments. Even heads of Wall Street firms begged LTCM to take their money. The hedge fund industry is founded on this belief. However, although some hedge fund managers have talent, most do not; quantitative analysis combined with qualitative analysis shows that most superficially spectacular track records can be explained by a combination of systematic risk exposures and luck.

Tips

Some of you will be reading this because you are contemplating a career in quantitative finance. The first job in finance is the hardest to find, and you might be looking for tips on how to get started. First, something which *doesn't* work: I receive countless resumes from job candidates who regard it as self-evident that academic success in some physical or mathematical science qualifies them as a finance quant. Apparently, they don't think any evidence of interest, understanding, or aptitude for finance is required. I feel like writing back, saying that although I have barely heard of their discipline nor ever read a book about it, I feel I am eminently qualified for a job in their university or research laboratory because of my understanding of finance! If you want to be more than a technician, you will have to learn some finance; finance is not a corollary (trivial or otherwise) of any mathematical science.

So what might work? My suggestion is to read and understand a couple of books. My favorites are Hull, *Options Futures and other Derivatives*[8] and, to get away from the derivatives pricing and CAPM rut into something more generally applicable, Luenberger's *Investment Science*. If these books seem hopelessly trivial to you, then try answering

Peter Carr's *FAQ's in Option Pricing Theory*.[9] Then, a tip given to me by Doug Steiner at RBC Dominion, *think up one money-making idea* and be prepared to defend it. This will generate a lot more interest at your interview than the Malliavin calculus!

This advice might get you started. But if you want to make your mark, find a really big financial problem and *create* a quantitative framework for dealing with it. You will need some imagination. Good luck.

Chapter 17

Steve Allen

Deputy Director, Masters Program in Mathematics in Finance, Courant Institute of Mathematical Sciences, New York University

How did I become a quant? I could give a cynical answer: By being lied to and being too naive to see through it. I could give an inspirational answer: By always letting my love of mathematics dictate my career choice. Both contain some element of the truth.

In Which the Author Is Seriously Misled

When I graduated from college in 1967, I had an education that was primarily focused on theoretical mathematics—particularly, mathematical logic. When I decided that I was not ready to start graduate school and needed to find a job, I really had no idea of what type of job I could do. So I went to an employment agency, asked what type of work involved use of mathematics, and was told that the ideal position for me would be with one of the new operations research departments being established by the New York banks, since they were utilizing "the most sophisticated mathematics of anyone other than NASA" (you can see that the term *rocket scientist* was just waiting to be born).

After a few interviews, I landed a job with the operations research department of Chase Manhattan Bank, whose interviewers had assured me that they were indeed applying advanced mathematics to the problems of finance.

Reality turned out to be a bit different. Although Chase (like most of the other large New York banks), had recently established a large operations research group, management really had very little interest in the projects it was proposing. In order to survive, the Chase group was basically taking on large programming jobs. In 1967, with the very first wave of computers reaching banking and well before there were any computer science majors, programming resources were hard to come by, and any group that could deliver completed systems, however inefficiently, was valued. And anyone whose background showed any hint of ability to learn how to program was considered a suitable hire. That's where I fit in.

I had never worked with computers before, but I figured I could pick up the knowledge I needed through courses. One week later, having been provided a set of IBM "Teach Yourself to Program COBOL" manuals, I was told I was ready to begin. I didn't feel very ready, but figured that, since there were six other members of the programming team I'd been assigned to, I could always ask questions of more experienced colleagues when I ran into trouble. I soon found out that nobody on the team had more than six months' experience. In fact, they were all coming to me with questions—after all, I was the one who had read the self-study manuals most recently.

During the first year I worked at Chase, all of the people with any experience in the operations research department left the bank, frustrated with management's lack of interest in their quantitative skills. When Chase showed no signs of ever replacing these experienced individuals, even the less experienced people started to exit, alarmed by the signs that the division was folding. From a group of 35 at the time I joined, the division rapidly shrank to just 5.

Why was I one of the few who stayed on amid this chaos? Partly because I was young enough and naïve enough to believe senior management's reassurances of turnaround that others were perceptive enough to see through as formulistic corporate cant. Partly because the departure of more senior people allowed me to take over management of the project I was working on, building a system for tracking bond portfolios, and I was finding that, through trial-and-error, I was learning to be a reasonably competent programmer. And partly because, while the project was primarily a large processing job, there was at least some quantitative content in terms of bond mathematics. Although I had been firmly instructed that bond mathematics was the province of the design team of the project, that my job was just to turn the equations into code, and that there was no need (or support) for me to learn the underlying mathematics, my interest in mathematics led me to take time on evenings and weekends to study this on my own. As a result, I was beginning to interact with the bond traders in suggesting changes in formulas, which gave my job enough quantitative content to feel rewarding.

With time, a new Chase management team began to perceive a role for quantitative analysis and eventually brought in a group of experienced professionals to head operations research. I was a junior member of the team, but by now had established a reputation of being able to manage a project, deliver results, and argue about formulas with clients of the group. So I was given a chance to work on projects as the group expanded its role into quantitative analysis of banking issues.

At first, the areas we were allowed to explore were more industrial than financial, such as a Monte Carlo simulation of check processing that was used to optimize a truck delivery routes from the branches to head office and check-sorting algorithms. Then we began to get involved with more financial applications—simulations of how the Bank's earnings would be impacted by different economic environments, which led to the measurement of risk and return tradeoffs for different asset-liability management strategies. By 1977, I had moved out of the operations research division to head efforts to build models and systems for a new asset-liability Committee and then to start a modeling group to support decision making on the firm's trading floor.

In Which a Fortuitous Opportunity Appears

This put me in good position to seize an opportunity when Chase began to consider the possibility of market making for foreign exchange and interest rate options in the early 1980s. By then, the impact of the Black-Scholes model on equity options in the late 1970s had begun to get some publicity and a few banks had begun applying the technique to foreign exchange and interest rates. But the supporting structure we are used to today—the textbooks, university courses, journals, and consulting firms specializing in options mathematics—were not yet in place.

From the information I was able to gather, the new field sounded very promising—a more serious use of mathematics in financial decision making than anything I had previously encountered. I knew I had handicaps in tackling this new field—it was largely built on parts of mathematics that were outside my primary fields of study. I had never even studied differential equations. But my orientation toward moving my career in the direction that promised the deepest possible involvement with math motivated me to take a chance. I began studying as hard as I could to position myself as the firm's "expert" in options theory. And it worked. I built up a group reporting to me that provided the modeling and analytic support to the firm's traders and structurers over the next 10 years.

This certainly proved to be the most pivotal moment in my career as a quant. Not only were the years I spent directing this team highly productive and enjoyable, but that experience is what established my credentials for my later career in risk management for derivative products and model review. What were the factors that allowed me to establish myself in a position for which my prior credentials seem, in retrospect, quite meager? One factor was clearly that it was early enough in the development of the field that there were few people available who did have strong credentials. So I enjoyed a *first-mover advantage* by staking my claim early. I could also draw on the reputation for delivering results I had established with key traders and managers. But I also believe I enjoyed an advantage in having to struggle to come up the curve mathematically. It put me in a good position to figure out how to educate others who were finding the new concepts difficult. Let me give an illustration.

The first options product for which Chase became a market maker was foreign exchange options. Senior trading managers were initially skeptical of this new product. Although they had heard some success stories of other firms in options trading, there were also some horror stories of significant losses and some strong opinions starting to circulate that the Black-Scholes theory had major theoretical weaknesses: It made unrealistic assumptions about lognormal distributions, continuous trading, and lack of transactions costs. Some people were claiming it was just a self-fulfilling prophecy—the only reason it worked at all as an indicator of prices was that other market participants were also using it.

In trying to get managers to overcome these fears, theoretical arguments from the mathematical derivations would be of little use—they didn't understand enough stochastic calculus to have any intuitive grasp of the arguments and were dubious about the relationship between the mathematics and market realities. Fortunately, I could draw on my own experiences in persuading myself that the theory worked.

Utilizing a database of closing daily FX rates over several years, I set up a simulation study of how FX options written for many different strikes, starting dates, and tenors would have fared using Black-Scholes delta hedging, rehedging only at the end of each day. The results showed that there was almost no dependence between profit and loss and where FX rates ended the day. There was a great deal of variance in P&L, but this could virtually all be attributed to whether realized volatility was higher or lower than the implied volatility at which we assumed the option would be priced. I could, therefore, use these simulation results to demonstrate that hedging based upon the Black-Scholes theory could produce reasonable results without assuming continuous trading and with a frequency of hedging that would not involve ruinous transactions costs. It showed that success of a market making operation in options would be a relatively simple matter of whether volatility could be priced at reasonable levels and should not be difficult to analyze. Furthermore, it was done in a way that could be easily demonstrated to traders who might not understand the mathematical theory but could certainly follow the details of a historically simulated set of trades.

In Which Reason Prevails and All Rejoice

Much has changed in the financial industry since those early days of options trading. New generations of traders and managers have entered the industry at a time when derivatives instruments were common and so have developed a degree of comfort with derivatives. Although they may not understand the details of the mathematics used in pricing derivatives, they have at least acquired a comfort level with the type of reasoning that underlies the mathematics. An important consequence of this is widespread acceptance of arbitrage-based reasoning for pricing instruments. This has resulted in a definite shift in the way the industry regards people with quantitative training.

When I moved from Chase's operations research division to its trading operation, I was the first analyst they had ever hired without an academic background in economics. The head of trading, in offering me the job, noted the limited role I would be able to play, since I would not be able to make my own forecasts of economic variables. But he expressed confidence that, as I worked in the field, I would begin to develop my own opinions of the economic future.

As things turned out, I didn't change but the environment did. The importance placed on making economic forecasts began to fade relative to the importance of understanding the right relationships between instruments. Mathematical modeling stopped being just an auxiliary function to work out the detailed consequences of an overall economic view and began to play a central role in developing new products and improving the pricing and hedging of existing products.

In the early 1980s, the decision-making process for pricing transactions lacked precise guidelines for determining probabilities attached to economic scenarios. In the absence of precise guidelines, the temptation was strong (and seldom resisted) for proponents of a particular action to tweak probabilities until they found values that supported their point of view.

For example, a team that had been working on putting together the structure of a complex investment had strong psychological and financial motivations to find arguments to support the desirability of going ahead with the investment. If an initial study showed the investment undesirable, a little sensitivity analysis could determine how the probability assumptions needed to be altered to support approval of the investment. It was easy for a motivated team to convince itself of the rightness of their new assumptions. Persuasion of other decision makers often hinged on internal political power. One consequence was that little value was placed on people with primarily quantitative skills—it didn't take much mathematical acumen to figure out how to change probability assumptions to get desired results.

The revolution in analysis that transformed the financial markets in the 1980s was the replacement of much of this subjective judgment about probabilities with actual market prices at which contingent cash flows (i.e., scenario-dependent cash flows) could be purchased and sold. This revolution consisted of two mutually reinforcing trends:

1. The growth of publicly traded markets in contingent claims such as options
2. The development of arbitrage-based analytic techniques for determining combinations of publicly traded contingent claims that could come close to reproducing complex scenario-dependent instruments

These two components were mutually reinforcing because the advances in analysis led to a greater demand for publicly traded contingent claims, as more users of complex instruments were motivated to use them for reducing uncertainty, while the availability of more publicly traded hedging instruments widened the scope of building blocks available for arbitrage analysis.

This significantly reduced the scope for political manipulation of results and expanded the influence of mathematically trained personnel who could provide the relatively objective arbitrage analysis that was now required. Initially, this demand for quantitative training was primarily met by hiring mathematics and physics PhDs. But as the demand continued to grow, many people in the industry began to question the efficiency of undergoing the years of specialized training required for a math or physics PhD before retooling for a financial career. Some schools began starting specialized master's programs in quantitative finance designed to offer a quicker route into the field to mathematically adept undergraduates.

I became involved with the program started at New York University's Courant Institute of Mathematical Sciences and, when I decided to retire from investment banking, I was able to transition to a full-time role in teaching and in helping to run that program. Over the past seven years, I have been gratified to witness the success of this and similar programs in preparing a new generation of quants for the industry.

In my new role, I frequently host question-and-answer sessions for potentially interested students. A frequent question is whether people joining the quantitative finance field now will have an opportunity to be creative, or whether the field is so well established that they will just be applying existing models in a routine fashion. My answer is that if you want to have a good career, you had better find a way to be creative; it's unlikely that your personality will be charming enough to induce an employer to pay you well for routine performance. But I add that creativity takes many forms—it consists not just of finding some new mathematical solution but also of discovering new ways to communicate results and build consensus.

We quants are fortunate to be living in an era that values our skills, but with rewards come responsibilities. We need to continue to find ways of making the insights of our craft available to people from many backgrounds and skill sets throughout the financial community.

Chapter 18

Mark Kritzman

President and CEO, Windham Capital Management, LLC

My life as a quant can be summed up by a Woody Allen quote that "90 percent of life is showing up." I did not become a quant by thoughtful design. Instead, early in my career, serendipity led me to a variety of situations that called for quantitative solutions. These experiences raised my level of interest in quantitative methods, and I endeavored to become more proficient. As I developed my quantitative skills, I discovered more opportunities to apply them and came to know many others in the field who shared my appreciation of quantitative methods.

A Brief Chronology

Upon graduation from college with a bachelor of science degree in economics and little exposure to quantitative methods aside from some introductory calculus and statistics, I accepted a position in the pension department at Equitable Life Assurance Society, then the third largest insurance company in the United States. My initial responsibilities involved performance calculations for pension funds—at this time, the state of the art in performance measurement was time-weighted rates of return. The notion of risk-adjusted returns had not yet gained traction at Equitable.

Reasonably soon after my arrival at Equitable, I managed to land a job in the Investment Advisory Department, which managed the asset allocation of Equitable's pension fund clients. This group determined how to allocate the funds across the various separate accounts that were invested in money market instruments, publicly traded bonds, direct placement bonds, large stocks, small stocks, and real estate. It was here that I first encountered mean-variance analysis, which upon reflection marks the beginning of my career as a quant. By this time, I had also begun pursuit of an MBA degree by studying evenings at New York University, so I was able to learn about mean-variance analysis both as a student and by experience. With the help of Chester Spatt,[1] then an undergraduate at Princeton who consulted with us, equitable developed one of the industry's first mean-variance asset allocation models, and we added a capability to perform risk analysis with assistance from NYU professors Aaron Tenenbein, Ned Elton, and Marty Gruber.

It was during this period that I made my first public presentation, at the Financial Analyst Federation's annual conference in Houston. I showed how to augment a corporation's income statement and balance sheet to account for the probable performance of the pension fund, taking into account the expected return and risk of the fund, net of the pension liabilities. Ironically, nearly 30 years later this is again a hot topic, as companies come to grips with their vulnerability to pension fund performance.

While at Equitable, I began to participate in a variety of industry forums such as the Institute for Quantitative Research in Finance (the Q Group) and the Investment Technology Association, which is now called the Society for Quantitative Analysis (SQA). One year, in the early 1980s, I was responsible for the SQA conference program and lined up Fischer Black, Eugene Fama, Bob Merton, Stew Myers, Myron Scholes, Steve Ross, and Jack Treynor as speakers, among others. I also attended the CRSP conferences at the University of Chicago and the Berkeley Program in Finance. It was at these gatherings that my interest in quantitative methods gained momentum.

In 1980, I accepted a position in the investment department of AT&T. At the time, AT&T was in the process of consolidating the pension funds of the regional telephone operating companies into a single fund to be administered centrally in New York. The consolidated AT&T pension fund, at $53 billion, became the largest fund in the world. Moreover, it had a total of 113 active asset managers and was the largest client for many, if not most, of these managers. As the in-house quant at AT&T, I was invited to a variety of industry events and became acquainted with many of the top quants throughout the industry and in academia. I also worked closely with the financial economics group at AT&T's research arm, Bell Laboratories—in particular, with Vijay Bawa of stochastic dominance fame and Stephen Brown, now at NYU's Stern School. Stephen and I subsequently coedited a book titled *Quantitative Methods for Financial Analysis*, for the CFA program. While at AT&T, I enrolled in the Ph.D. program at NYU's business school and began to hone my quantitative skills. After a couple of years I transferred to the graduate school of arts and science at NYU to pursue a Ph.D. in economics. I soon realized that I could not continue with my career, especially given my travel commitments, and at the same time devote the requisite attention to my classes. Reluctantly, I abandoned my formal academic training.

From AT&T, I eventually moved to Bankers Trust, where I focused on the development of a risk model, with the help of Bob Litzenberger, then at Stanford University, and Krishna Ramaswamy at Wharton, and I became interested in portfolio insurance. My colleague at Bankers Trust, Richard Tanenbaum, founder of Savvysoft, led our portfolio insurance initiative.

Having worked at three large institutions, I decided to try something more entrepreneurial and cofounded New Amsterdam Partners with Tony Estep and Michelle Clayman, who were both equity quants at

Salomon Brothers. I am happy to say that New Amsterdam Partners continues to prosper with Michelle at the helm. I left New Amsterdam Partners after a couple of years to start Windham Capital Management with Neil McCarthy.

Neil and I eventually split the business into a New York company, which focused on U.S. equity management, and a company located in Cambridge, Massachusetts, which focused on currency management. The New York company was eventually purchased by Oppenheimer, while Windham Capital Management, LLC in Cambridge continues to this day as a currency manager, with more than $30 billion of assets under management, which we manage jointly with State Street Bank. Windham also provides portfolio and risk-management software and consulting services to institutional investors throughout the world.

With this chronology as a backdrop, let me now turn to how I developed my quantitative skills.

How I Developed My Quant Skills

Three experiences come to mind as the catalyst for my development as a quant. First, my positions with large, influential institutions, especially AT&T, enabled me to develop acquaintances and, in some instances, friendships with some of the industry's leading quantitative lights. Much of what I learned about portfolio theory and asset pricing resulted from one-on-one conversations at conferences and other events, or by observing and participating in discussions among the likes of Bill Sharpe, Harry Markowitz, Bob Merton, Steve Ross, Jack Treynor, and Fischer Black. Had I not worked at these institutions, I probably would not have gained the direct access I had to these people.

The second catalyst arose from my experience as an entrepreneur. During the startup phase of my business ventures, I lacked resources to purchase software for performing various quantitative tasks or to hire consultants. Therefore, I was forced to develop many applications myself. It's one thing to understand a quantitative application conceptually. It's quite another thing to develop the application yourself.

The third catalyst for my development as a quant was my experience as a contributing editor for the *Financial Analyst Journal*. For several years

I wrote a column titled "What Practitioners Need to Know." For each issue I chose a topic, usually a quantitative one, and tried to present it with a reasonable amount of rigor without resorting to unnecessary symbolism and jargon. I found that the challenge of explaining a topic to others who lacked serious quantitative training was an excellent means by which to gain a thorough understanding of the topic. As I quickly exhausted topics with which I was reasonably familiar, I found myself learning about new topics, which helped me to expand my quantitative skills. It was through these experiences rather than in the classroom that I learned most of what I now know about quantitative methods.

How I Applied My Quantitative Training

I think of a successful quant not as someone who is merely mathematically proficient, but as someone who asks the right question and then applies the appropriate quantitative tools to answer it. Here are some examples of how I attempted to respond to interesting questions with quantitative tools. The first few examples deal with innovations in portfolio and risk management. The second set of examples focuses on dispelling myths.

Innovations

Surplus Insurance. In the mid-1980s, portfolio insurance was the rage. I recognized this as a clever and useful development, but I thought its application to pension funds was misguided. Why insure the absolute value of the fund? Why not insure the pension plan's surplus instead? It was this value, after all, that drove the cost to the plan sponsor and determined the solvency of the pension plan. I had recently read Bill Margrabe's article about valuing an option to exchange one risky asset for another[2] and immediately recognized this valuation framework as the solution for insuring a pension fund's surplus. I argued in a 1986 article in the *Journal of Portfolio Management*[3] that surplus insurance should be less expensive than portfolio insurance, because the net volatility of assets and liabilities would likely be lower than the volatility of the assets by themselves. I also argued that insuring the assets might perversely increase the volatility of the surplus, because the insured assets would be less correlated with the liabilities than the uninsured assets. Of course, the stock market soon thereafter crashed, in part, as a consequence of portfolio insurance. If pension fund sponsors had insured their surpluses rather than their assets, I wonder how differently things might have turned out.

Within-Horizon Risk. Several years ago, I met with the investment staff of a foundation to help them determine their asset mix policy. They stated that their objective was to identify an asset mix policy that, in combination with their spending plan, would have a high probability of exceeding a particular loss threshold at the end of 20 years. It occurred to me that they might also care about what happened in the meantime—say after one year, two years, or five years.

After some discussion, they acknowledged that this threshold was relevant not just at the end of a 20-year horizon but throughout their investment horizon. I simulated their exposure to loss, assuming the fund was monitored throughout the horizon and discovered it was considerably higher than most people would expect. I mentioned this to Don Rich, a professor at Northeastern University at the time, and he claimed that we could develop an analytical solution to this problem using first passage time probabilities.

This work led us to develop a new measure of value at risk called *continuous value at risk,* which gives the worst outcome at a particular probability assuming the investment is monitored continuously. The paper we wrote about within-horizon risk won a Graham and Dodd Award, and our within-horizon risk measures are now widely used throughout the industry.[4]

Full-Scale Optimization. A couple of years ago, I wrote a short piece for Peter Bernstein in which I showed the mathematical equivalence of portable alpha strategies with currency overlay strategies. Soon after publication, I received a note from Paul Samuelson critiquing my approach and mean-variance analysis in general. He argued that computational power now enables us to maximize plausible utility functions numerically, based on the entire distribution of returns, not just the first two moments. He suggested that I try his approach and compare the differences, which of course I did.

I discovered that the mean-variance approximation performed quite well for variations of power utility, even with significantly nonnormal return distributions. However, utility functions that incorporated kinks or inflection points proved problematic for mean-variance analysis. Given the growing popularity of hedge funds and the concern for thresholds that many investors had, I worked with my colleagues, Tim Adler, Jan-Hein Cremers, and Sébastien Page to develop a robust full-scale optimization algorithm.[5] As a consequence of Paul Samuelson's critique and my eagerness to understand it, we introduced an entirely new way of constructing portfolios that easily addresses nonnormality and a wide range of utility functions.

Efficient Trading. Institutional investors often reallocate their portfolios to shift their asset mixes or to shuffle their investment managers. Often they delegate the execution of these portfolio reallocations to transition specialists who are positioned to trade cost-effectively, usually because they can cross trades internally. The most significant risk faced by transition managers is opportunity cost, which arises when the legacy portfolio falls in price before it is sold or the target portfolio rises in price before it is purchased. Transition managers try to control this risk by minimizing the differences in sector weights between the legacy and target portfolios as the transition unfolds.

It occurred to me that one could improve upon this approach by modifying an algorithm originally introduced by Bill Sharpe for portfolio optimization.[6] Sharpe showed that one could identify the utility-maximizing portfolio by taking the partial derivatives of expected utility with respect to the asset weights, and then continually shifting the portfolio from the asset with the lowest derivative to the asset with the highest derivative until all the derivatives were equal. His intent was to pose an alternative to large-scale matrix inversion, which at the time was computationally challenging. As computers became more powerful, Sharpe's algorithm became less relevant to portfolio optimization.

Notwithstanding the obsolescence of Sharpe's algorithm applied to portfolio optimization, Simon Myrgren, Sébastien Page, and I found a way to apply a variant of Sharpe's algorithm to the portfolio transition problem. We modified Sharpe's algorithm to focus on tracking error rather than expected utility and to conform to the constraint that purchases and sales must be offsetting to ensure that the transition is self-financing. Tests of our algorithm showed that it reduced exposure to opportunity cost by 40 percent compared to the prevailing industry approach.[7]

Dispelling Myths

Time Diversification. Paul A. Samuelson famously demonstrated that time does not diversify risk under commonly invoked assumptions about investor preferences and investment returns.[8] Nonetheless, many academics and most practitioners failed to grasp his point and stubbornly tried to demonstrate that time does diversify risk. In one of the cleverest articles I have ever read, he responded to his critics by making his point using words of only one syllable except the last word, which was *syllable*.[9]

I, too, was confused about the topic, so I wrote to him seeking clarification, and he responded with an explanation that inspired me to present his argument in a way that I thought would appeal to nonquants.[10] I calculated the certainty equivalent of a risky gamble for a logwealth investor and used a binomial tree to illustrate the equivalence. Specifically, I showed that a $100 investment with an equal chance of a one-third gain and a quarter loss produced the same expected utility as $100 for sure. I then extended the binomial tree an additional period and showed that although expected return increased, expected utility remained constant. I used this pedagogical construct to expand the discussion to alternative specifications of expected utility and nonrandom returns. I believe my pedagogical approach helped many nonquants grasp the fallacy of time diversification.

Are Optimizers Error Maximizers? Hype versus Reality. In my business and research, I often apply mean-variance optimization, and for the most part obtain results that are robust to reasonable input errors. Nonetheless, I often hear the refrain that small errors in the inputs to a mean-variance optimizer lead to large errors in its output. I am not sure how this belief became so prevalent, but I have a couple of conjectures.

Optimizers were first used to allocate portfolios across marketable securities such as publicly traded stocks and bonds, and they performed quite reasonably. Then investors included real estate and privately placed bonds. The reported returns of these asset classes were based on appraisals and matrix pricing rather than market transactions; hence, they displayed artificially low volatility. When these asset classes were introduced to the optimizer it indicated that most of the portfolio should be allocated to them. Moreover, it showed that such an extreme allocation would substantially improve the risk/return trade-off. My conjecture is that critics of optimization latched on to this result to hype the sensitivity of optimizers to input errors. Of course, informed users of optimization understand the problem and employ a variety of methods to adjust the volatility assumptions appropriately, and thereby obtain reasonable results.

My second conjecture is more cynical. Some quants have devoted a great deal of effort and resources to developing methods to ameliorate estimation error and are therefore vested in the notion that small input errors result in large output errors.

Given that my experience belies the conventional wisdom, I thought it might be useful to calibrate the sensitivity of mean-variance optimization to input errors with a couple of examples that typify the use of optimizers. In the first example, I optimized across a sample of country equity indexes. I estimated the standard deviations and correlations from the available set of monthly returns, and I used the historical covariance matrix to calculate equilibrium returns. In this example, the expected returns, standard deviations, and correlations were not particularly dissimilar. I then identified the efficient portfolio given a particular risk-aversion coefficient. I next conducted a second optimization in which I assumed I overestimated the means of half the assets by 1 percent and underestimated the means of the other half by the same amount. I again identified the efficient portfolio based on these erroneous inputs.

My results showed that the error-determined portfolio was about 60 percent misallocated. This large misallocation occurred because the assets were close substitutes. But because the assets were close substitutes, the misallocation didn't much matter. The correct portfolio and the incorrect portfolio had very similar return distributions; hence, their exposure to loss and likelihood of gain were not far apart. I then repeated the experiment, but this time for a set of assets that were significantly dissimilar. In the second example, the 1 percent errors in the means had little impact on the weights and the probability distributions. These experiments showed that mean-variance optimization, at least in two very typical applications, is robust to estimation error when measured appropriately.[11]

The Hierarchy of Investment Choice. Peter Bernstein asked me to tackle the question of the relative importance of asset allocation and security selection, so I did. It occurred to me that the obvious way to address the issue was to simulate returns by holding fixed one decision and varying the other. Then I could determine which decision generated more dispersion in wealth. This experiment would allow me to measure the dispersion in performance that arises naturally by engaging in a particular investment activity. Other approaches for sorting out the relative importance of asset allocation and security selection, such as Brinson, Hood, and Beebower (BHB)[12] and Ibbotson and Kaplan[13] focused on the realized returns of managed portfolios; consequently, these studies failed to disentangle investment behavior from investment opportunity. My colleague, Sébastien Page, and I performed the analysis using bootstrap simulations of available returns and discovered that security selection was overwhelmingly more important than asset allocation.[14]

This outcome, which we did not anticipate, provoked considerable debate among academics and practitioners. As this debate intensified, I reexamined the BHB methodology and discovered it was specious for reasons other than its reliance on realized returns. I contrived an experiment to demonstrate its fundamental flaw. I hypothesized a world in which all asset classes had the same performance, but within each asset class the performance of individual securities varied significantly. In this hypothetical world, security selection explained 100 percent of the difference in the performance among funds, while asset allocation had no impact whatsoever. I essentially created a world with a single asset class, thus rendering the asset allocation decision irrelevant. I then applied the BHB methodology, and it revealed that asset allocation determined 100 percent of performance and security selection determined none of it—the exact opposite of the truth.[15]

The Future for Quants

Quantitative analysis has advanced from the fringes of the investment management profession to the mainstream and is well on the way to becoming the dominant paradigm of the investment industry. Owing to its rise in popularity, however, mathematical proficiency will not be sufficient to guarantee a successful career as a quant, especially as these skills become more commoditized. The successful quant will combine mathematical proficiency with an appreciation for economic and financial theory, and he or she must know which questions are really important.

Chapter 19

Bruce I. Jacobs and Kenneth N. Levy

Principals, Jacobs Levy Equity Management

Our adventures in quantitative equity have been a joint endeavor since 1986, when we cofounded Jacobs Levy Equity Management, now a $20 billion institutional asset management firm. Even before then, however, our separate paths seemed destined to converge. Perhaps it was inevitable, given our mutual interest in quantitative finance and the narrowness of the quantitative equity field at the time. In any event, our story begins with two separate voices that eventually merge into one.

Portraits of Two Investors

Bruce: As a teenager interested in the stock market, I convinced my parents to let me open a brokerage account to test my own method of investing. I read mutual fund reports to identify which stocks the funds were buying or selling in common. I then did some fundamental and technical analysis on these names and bought two shares each of six different companies to diversify my portfolio. I didn't know it at the time, of course, but this little hobby would ultimately develop into a career.

At Columbia College, I decided to enroll in its three-two program, which meant that I spent three years studying the contemporary civilization and humanities core curriculum, as well as the hard sciences, and then two years at the Columbia School of Engineering. There, I found a home in operations research, which allowed me to study computer science and applied mathematics, including differential equations, stochastic processes, statistical quality control, and mathematical programming. While studying for my master's in operations research at Columbia, I had the opportunity to work at the Rand Institute, where math and computer science were applied to real-world problems. There I was involved in developing a large-scale simulation model designed to optimize response times for the New York City Fire Department.

My interest in applied math led me to Carnegie-Mellon's Graduate School of Industrial Administration, which had a strong operations research faculty. There I studied applications of management sciences in accounting, finance, marketing, and production. I quickly became enthralled with finance, given its mathematical content and emphasis on economic decision making over time and under uncertainty. I earned an MBA at Carnegie-Mellon, and went on to further my graduate education in finance at the Wharton School of the University of Pennsylvania. I eventually earned my master's and PhD there and served on the finance faculty for several years, teaching both undergraduates and MBA candidates. Little did I know that one person in the PhD program was to become a very important part of my future in quantitative equity.

Ken: I followed a different path. When I was in my teens, I began investing my earnings from summer jobs in individual stocks, basing investment decisions on fundamental data provided by my father's broker and on my own handmade price charts. While the stock market moved sideways for the next 15 years, my interest in the market continued to grow.

I had always assumed I would end up working in my family's wholesale distribution business. With that future in mind, I focused my education at Cornell on economics and related liberal arts disciplines. I had a strong quantitative aptitude, however, and my favorite courses were the computer science and operations research offerings from the engineering college. I went on to earn an MBA from the Wharton School, with a major in general management.

MBA in hand, I joined the family business, using my quantitative skills and knowledge to design systems for sales forecasting and inventory control, writing computer code for the entire operation. (This was in the days before off-the-shelf software for these tasks was available at the local office supply store.) After five years of this, I was out of challenges and needed a change. My father—who was also my boss—encouraged me to pursue my passion for finance and the stock market.

I enrolled in the doctoral program at Wharton, where I earned a master's upon completing the coursework and qualifying exams for the doctoral program. At that point, I decided I had gained enough skills to test the waters back in the real world. The Wharton School's job placement center had little experience placing PhD students in the financial industry. One of my professors, however, noted that another Wharton faculty member, Bruce Jacobs, was exploring a similar path.

New Concepts, Foggy Ideas

Bruce: I remember coming home from my first semester at Wharton and being asked by my father, "What did you learn so far?" I responded, "I've learned three things, and this is all there is to know about finance. First, there's something called the *efficient market hypothesis,* which says that the markets are efficient and it's impossible for an investor to outperform the market. Second, there's something called the *capital asset pricing model,* which says that all you need to know about stocks to be an investor is a stock's beta, its sensitivity to market moves. Third, there's something called *Modigliani-Miller,* which says that the choice of a firm's capital structure, its debt to equity ratio, doesn't matter." My father then asked me what I was going to do with this knowledge. "I haven't the foggiest idea," I said.

The efficient market hypothesis was, of course, all the rage in academia at the time. Way back in the 1930s, Benjamin Graham and David Dodd began to systematize security valuation, spawning the thought that investing had more in common with science than with the local numbers racket. By the 1950s, Harry Markowitz was turning portfolio construction into a disciplined endeavor. The academic scene exploded in the 1960s with seminal ideas like the capital asset pricing model, and in the 1970s with arbitrage pricing theory and the Black-Scholes-Merton option pricing formula.

In 1965, the University of Chicago's Eugene Fama published "The Behavior of Stock Prices," which laid the foundation of the efficient market hypothesis. Fama theorized that stock prices fully and instantaneously reflect all available information. In the same year, Paul Samuelson at MIT published his "Proof that Properly Anticipated Prices Fluctuate Randomly," which showed that, in an efficient market, price changes are random and thus inherently unpredictable. Burton Malkiel at Princeton later popularized these views in *A Random Walk Down Wall Street*, published in 1973.

Academic analyses of the burgeoning amount of available data seemed to support market efficiency. Computer-enabled dissections of actual market prices suggested that price changes followed a random walk. Furthermore, Michael Jensen, one of Fama's doctoral students, analyzed mutual fund performance from 1945 to 1964 and found that professional managers had not outperformed the market.

If one could not predict security prices, active management was futile. The solution seemed to be to shift the emphasis from security selection to constructing portfolios that offered the market's return with the market's risk. Requiring no security research and little trading, these portfolios could capture the long-term upward trend in overall stock prices. Low-cost, passive index funds were born.

The efficient market hypothesis prevailed in academia during the 1970s, and Wharton was no exception. Many of the school's faculty held doctorates from the University of Chicago and had been students of Fama. They tended to discourage doctoral theses that contradicted the theory. I cast my own thesis in efficient market terms. (Much later, I realized that this thesis constituted the beginnings of my thinking on the earnings accrual anomaly, which would be modeled and put to work at Jacobs Levy Equity Management.)

The prevalence of this ivory tower thinking made it difficult for quants to find jobs on Wall Street in the early 1980s. This was many years before Fischer Black and other "rocket scientists" became fixtures on the scene, and the Street didn't know what to do with PhDs. Security analysis was still largely the realm of fundamental analysts parsing accounting reports and visiting companies. Fortunately, Prudential Insurance Company was willing to hire quants in asset management.

I did not know that, after leaving Wharton, Ken had landed a job in the equity management department at Prudential. We quickly reconnected. In our early days at Prudential, we found the organization as a whole reluctant to use quantitative methods and averse to innovation. Over time, there was more tolerance and, eventually, even support. I was able to carve out a new Prudential affiliate focused solely on quantitative investing, and this gave Ken and me the opportunity to work together.

Bruce and Ken: In this affiliate, we used commercially available tools, such as those provided by Barra, to construct equity portfolios. At the time, most of these tools were directed toward risk management. Portfolio risk management is a critical aspect of consistent performance, and one we have emphasized a great deal in our own work. It does not address, however, the burning issue of how to identify securities that will outperform in the first place.

Neither of us believed in the efficient market hypothesis and the impossibility of superior performance. We knew the power of quantitative methods. We were familiar with the statistical tools needed to analyze security prices and market behavior. And by the early 1980s, cracks were beginning to develop in the wall of market efficiency, cracks that hinted at the promise of superior performance.

New academic studies showed that certain types of stocks did perform better than the market average. Higher returns seemed to accrue to firms with smaller-than-average market capitalizations. Analysts tend to neglect such stocks, compared with larger-cap securities, and neglected stocks also tended to outperform.

Stocks with lower price/earnings ratios were found to perform better than stocks with higher price/earnings ratios, while stocks with lower price/book ratios performed better than those with higher price/book ratios. In some studies, low price itself seemed to herald high returns.

Empirical evidence indicated that stocks whose earnings estimates had recently been upgraded by analysts tended to produce above-average returns, perhaps because of behavioral reasons such as analysts' tendency to herd or their aversion to making substantial revisions in estimates. Researchers also found that earnings surprises tended to produce excess returns, and that negative surprises had a greater effect on stocks with high expected earnings growth than on those with low expected earnings growth.

These findings were anomalies within the context of the efficient market hypothesis. Not only did they suggest that beta alone was insufficient to understand stock returns, they also indicated patterns of stock price behavior that investors could have exploited to earn above-average returns. But if profits were to be had simply by buying low-price/earnings stocks or small-capitalization stocks, why weren't smart investors able to perform better than the market on a consistent basis?

We founded Jacobs Levy Equity Management in 1986 because we thought we had some unique answers to this question and might be able to develop the means to take advantage of our insights for the benefit of clients. To accomplish this, we needed an environment that would be conducive to the type of dedicated, deep research we had previously done at Wharton. This required giving up our responsibilities at Prudential, and our incomes.

The Jacobs Levy Investment Approach

Our investment approach is based on a philosophy of *market complexity*. We believe the equity market is not simple or ordered in such a way that a simple rule such as "buy low price/earnings stocks" or "buy small-cap stocks" delivers consistent profits; nor is the market totally random, hence unpredictable. Rather, equity market returns are driven by complex combinations of company fundamentals, economic conditions, and behavioral factors.

We believe security prices respond to numerous fundamental factors, including price/earnings ratios, expected growth rates, and analysts' earnings estimates, and to economic factors such as interest rates. Prices also respond to behavioral elements such as investors' tendencies to overreact to news, their desire to seek safety in numbers, and their selective memories. It is possible to detect these responses and to design stock selection models that can exploit them in order to deliver superior returns.

Doing so is not easy, however. Return-predictor relationships are likely to differ across different types of stocks. Because there are more financial firms among value stocks than growth stocks, for example, value stocks can be more sensitive than growth stocks to changes in interest rate spreads. Earnings estimate revisions and earnings surprises, by contrast, are more important for growth than for value stocks. So Google shares can take a nosedive when earnings come in a penny under expectations, while Bank of America shares are hardly affected by such a disappointment.

Once modeled, return-predictor relationships are likely to change over time. The world is constantly evolving, and old inefficiencies can disappear, giving way to new ones. Merely tilting a portfolio toward historical anomalies does not produce consistent performance. It takes ongoing research on new inefficiencies, new sources of data, and new statistical techniques to keep an investment approach in synch with evolving opportunities.

Finally, return-predictors are often correlated with each other. Small-cap stocks tend to have low price/earnings ratios, and low price/earnings ratios are correlated with high yield. Also, certain attributes may be correlated with industries. A simple high-yield screen will select a large number of bank and utility stocks. Such correlations can distort naïve attempts to relate returns to potentially relevant variables. Our seminal insight was *disentangling*: by modeling numerous potential return-predictor relationships in a way that takes interrelationships into account, a more accurate picture of the return predictors emerges.

Benefits of Disentangling

Perhaps we should have called our firm Jacobs Levy Equity Research, since we spent the first three years alone, doing just research, in what a future client referred to as a "Class D" office building. We were asked: "How are you two living? You don't have any assets under management yet and are too busy writing articles to do any marketing." "Not a problem," Bruce replied, "Ken pays my salary and I pay his." We had only limited access to a DEC VAX cluster, so our programs often ran all night on our own slow PCs. We made Saturday runs to the post office, waiting anxiously for a reply from the *Financial Analysts Journal*, where we had submitted our first article. In those early years, every call we received was from our wives, and the mail was easy to open—there was none.

Nevertheless, we persisted with our research. Our aim was to investigate *all* the market inefficiencies in the literature, and to uncover new ones. Other researchers were looking at one effect at a time, at most two or three. No one else was looking at all the effects simultaneously, to discover which ones survived in a multivariate setting.

The standard approach for measuring a return effect at that time entailed grouping a universe of stocks by, say, their price/earnings ratios and calculating the average return to the quintile or decile of stocks with the lowest ratios. Comparing this average with the average for the entire universe yielded a measure of the low-price/earnings effect. But this simple approach fails to account for correlations between return predictors or for the possible effects of industry affiliations.

We went way beyond the standard approach, disentangling return-predictor relationships via a simultaneous analysis of a multitude of relevant effects. With a multidimensional, simultaneous analysis, returns to each equity characteristic are purified by neutralizing the impact of all other measured effects. For example, the *pure* payoff to low price/earnings is disentangled from returns associated with related attributes such as high yield. Conceptually, the pure return to low price/earnings accrues to a portfolio that has lower-than-average price/earnings but is marketlike in all other respects; that is, it has the same industry weights, average yield, capitalization, and so on as the market.

Several aspects of our research surprised us. The research itself turned out to be far more difficult than we initially imagined, taking years rather than months. Looking back, of course, that's not so surprising. We were dealing with very complex issues, examining dozens of attributes of companies, investors, and the economy, and thousands of stocks in different market environments. And we were the first in the world to do so.

We were pleasantly surprised by the strength of our findings. In short, they strongly contradicted the efficient market hypothesis. We found the stock market rife with inefficiencies, consistent with our belief that investment opportunities can be detected and exploited to offer superior performance. As our philosophy and some of our findings began to be publicized, other quantitatively oriented money managers asked us to consult or to sell them our proprietary databases and disentangling code. We remained steadfastly committed to developing our own money management business. (Some years later, quantitative consulting firms such as Barra and Vestek began to develop products that managers could use to test various return predictors.)

Gaining our first clients was not an easy task. The eventual publication of our research helped a great deal, as it elicited a lot of interest from the investment community. What didn't help was that, in the early 1990s, quants were tertiary managers, considered by potential clients only after several fundamental managers were in place. Nor could we count on consultants to recommend our strategies, because they typically required three-year live track records. We finally found a few courageous pension officers willing to take some maverick risk for the benefit of their plans. We remain deeply grateful to these clients and are delighted that many of them are still our clients today.

We have been not exactly surprised but certainly heartened by the continuing robustness and success of our research. Over the past 20 years, Jacobs Levy has grown from a two-man research effort into a strong team of 60 with top industry talent in all functional areas. The firm has earned a spot on *Pensions & Investments*' list of Top 25 Managers of Active Domestic [U.S.] Equity, managing more than $20 billion for an international roster of over 50 clients. These include many of the world's largest and most sophisticated corporate pension plans, public retirement systems, multiemployer funds, endowments, and foundations.

The firm's success reflects the trust our clients have placed in us, and their trust in turn reflects the ability of our research, beginning with disentangling, to deliver value added. Disentangling distinguishes real effects from mere proxies, real investment opportunities from spurious ones. For example, the small-firm effect, measured naïvely, arises from a bundle of related attributes. Our research has shown that the January small-firm seasonal effect vanishes when disentangled from related effects; it proves to be a mere proxy for year-end tax-loss selling. As not all small firms will benefit from a January rebound, indiscriminately buying small firms at the turn of the year is not the best approach.

Disentangling reveals the true nature of the various return-predictor relationships. For example, stocks with low price/earnings are usually considered defensive. But pure returns to low price/earnings perform no differently in down markets than in up markets. The defensiveness of low price/earnings in naïve form arises because it is a proxy for defensive attributes such as high yield and defensive industries such as utilities. Disentangling can also reveal hidden opportunities. Small-cap stocks, for instance, may be characterized by low price and analyst neglect, as well as capitalization. Only a multivariate analysis can distinguish the extent to which returns accrue to each of these characteristics separately. Also, the pure returns that result from disentangling are additive. If analysis shows that positive returns accrue to both small capitalization and analyst neglect, the investor may benefit from both attributes by investing in small-cap stocks that are covered by relatively few analysts.

Pure returns also tend to be much less volatile than their naïve counterparts, because they capture more signal and less noise. Consider a naïve analysis of returns to low price/book. As most utilities have low-price/book ratios, a naïve return to low price/book will be affected by events such as oil-price shocks, which are relevant to the pricing of utility stocks but not necessarily to the pricing of other stocks with low-price/book ratios. By contrast, a pure return to price/book controls for the noise introduced by industry-related effects. By providing a clearer picture of the precise relationships between stock price behavior, company fundamentals, and economic conditions, disentangling improves return predictability.

We were delighted with the richness of our findings and hopeful that the *Financial Analysts Journal* would have an interest in them. In 1988, the journal published our paper, "Disentangling Equity Return Regularities: New Insights and Investment Opportunities," which introduced the concept of disentangling. This article won a Graham and Dodd Award as one of the best articles of 1988 and was subsequently translated into Japanese for the *Security Analysts Journal of Japan. Financial Analysts Journal* went on to publish "On the Value of 'Value'," "Calendar Anomalies," and "Forecasting the Size Effect." The *Journal of Portfolio Management* published "The Complexity of the Stock Market," introducing the notion of market complexity, in 1989. Editors Peter Bernstein and Frank Fabozzi later selected the article for the collection *Streetwise: The Best of The Journal of Portfolio Management*.

University of Washington Professor Charles D'Ambrosio, editor of the *Financial Analysts Journal* at that time, noted in the *Wall Street Journal* ("How Jacobs and Levy Crunch Stocks for Buying—and Selling," March 20, 1991) that we were "the first to bring so much of this anomaly material together." At his invitation, we presented our findings on complexity and disentangling at the CFA Institute's 1988 conference on continuing education. We also later presented them to the Institute for Quantitative Research in Finance ("Q Group").

Integrating the Investment Process

Our research laid the groundwork for our investment approach. Statistical modeling and disentangling of a wide range of stocks and numerous fundamental, behavioral, and economic factors results in a multidimensional security selection system capable of maximizing the number of insights that can be exploited while capturing the intricacies of stock price behavior. This, in turn, allows for construction of portfolios that can achieve consistency of performance through numerous exposures to a large number of precisely defined profit opportunities.

To preserve the insights gained from our security selection models, we realized from the beginning that we would need to build our own tools to implement those insights. We have developed customized, quantitative systems not only for security selection but also for portfolio construction, compliance, trading, and performance attribution. Integrating every step of the investment process across the same proprietary factors helps to ensure that the portfolio construction process fully exploits all detected investment opportunities and controls for all known risk exposures. Furthermore, with an integrated process, actual portfolio results can be used to evaluate security selection and provide input to the research process.

Insights can also be eroded by transaction costs, but we hold several advantages in the trading arena. First, because of our disentangling approach, we can profit from multiple inefficiencies for each security that we trade. Second, with our integrated systems, transaction costs are estimated and fed back to the portfolio construction process, helping to ensure that only economical trades are made. Third, we were early advocates and users of low-cost electronic trading venues. Finally, we maintain strict capacity limits to ensure that our trading remains nimble and cost effective. In 2006, *Institutional Investor*'s ranking of investment managers cited us as having the lowest costs for NYSE trading and the third lowest for Nasdaq trading.

Our thoughts on the importance of unifying the investment approach and integrating the investment process are outlined in two articles published in 1995, "Engineering Portfolios: A Unified Approach" (*Journal of Investing*) and "The Law of One Alpha" (*Journal of Portfolio Management*).

One of the great advantages of a quantitative approach is that it allows us to follow a very large universe of securities—virtually every U.S. stock with sufficient information flow and liquidity for institutional investors—and a multitude of attributes. We look at firm and market-based attributes such as earnings, accruals, value, growth, size, momentum, price reversals, and volatility; managements' informed actions and analysts' influential opinions; investor sentiment and other behavioral effects including investor underreaction and overreaction; industry affiliations; and a number of economic factors. This provides a basis for constructing portfolios that can meet a variety of client needs.

When we began our research process, we expected to offer portfolios that contained the best stocks according to our stock selection system; the client could measure the portfolio's return without regard to any particular benchmark. But as our research progressed, new market indices were being developed based on capitalization (large, mid, and small) and style (growth and value). Consultants began advancing the notion of constructing portfolios relative to given indices so that clients could better benchmark manager performance. The breadth of our investment universe and our customized, quantitative portfolio construction methods have allowed us to design portfolios for any number of mandates; new portfolios tied to new underlying indices emerge on a regular basis.

Relaxing Portfolio Constraints

We were acutely aware of the costs associated with constraints on portfolio construction, including constraints on the selection universe and on risk-taking. In the early 1990s, we found a client willing to relax the universe constraint and fund a full-universe portfolio. Later, we resisted the push toward *enhanced indexing,* with its tightly controlled residual risk limits. Our 1996 article, "Residual Risk: How Much Is Too Much?" (*Journal of Portfolio Management*), delineated the advantages of being more opportunistic with respect to residual risk taking.

We were also fully aware of the cost of constraints on short selling, but we did not think short selling would be acceptable to pension fund clients. Soon after we began managing portfolios, however, some clients asked about shorting stocks. With their encouragement, we ran analyses on the stocks at the bottom of our return prediction rankings and found that they did underperform the market.

Jacobs Levy soon became one of the first money managers to exploit the potential of short selling within a disciplined framework when we began offering long-short portfolios in 1990. Engineered long-short portfolios offer the benefits of shorting within the risk-controlled environment of quantitative portfolio construction. The ability to sell stocks short can benefit both security selection and portfolio construction. To begin with, short selling expands the list of implementable ideas to include both "winning" and "losing" securities. Portfolios that cannot sell short are restricted in their ability to incorporate insights about losing securities. For example, a long-only portfolio can sell a loser if it happens to hold one, or it can refrain from buying a loser. In either case, the potential impact on portfolio return is limited by the absolute weight of the security in the benchmark.

Consider that the typical stock in a broad market index such as the Russell 3000 constitutes about 0.01 percent (one basis point) of that index's capitalization. Not holding the stock (or selling it from a portfolio) gives the portfolio a 0.01 percent underweight in the stock, relative to the underlying benchmark index. This is unlikely to give the portfolio's return much of a boost over the benchmark, even if the stock does perform poorly. It also does not give the manager much leeway to distinguish between degrees of negative opinions; a stock about which the manager holds an extremely negative view is likely to have roughly the same underweight as a stock about which the manager holds only a mildly negative view.

Short selling removes this constraint on underweighting. Significant stock underweights can be established as easily as stock overweights. The ability to short thus enhances the manager's ability to implement all the insights from the investment process, insights about potential losers as well as winners.

Short selling also improves the ability to control risk. Benchmark weights are the starting point for determining a long-only portfolio's residual risk. Departures from benchmark weights introduce residual risk, so a long-only portfolio tends to converge toward the weights of the stocks in its underlying benchmark in order to control risk. The need to converge toward benchmark weights necessarily limits the portfolio's potential for excess return, as returns in excess of benchmark accrue only to positions that are overweighted or underweighted relative to their benchmark weights. In a portfolio that can sell securities short, the risks of the securities held long can be offset in part or in full by the risks of the securities sold short.

We described the benefits of shorting and long-short portfolios in several articles, the earliest being "Long/Short Equity Investing," which appeared in the *Journal of Portfolio Management* in 1993 and was later translated into Japanese for the *Security Analysts Journal of Japan*. This was followed by "20 Myths About Long-Short" (*Financial Analysts Journal*, 1996) and other articles. We presented our long-short research to the CFA Institute's 1993 and 1998 conferences on continuing education and to the Q-Group in 1995.

Short selling can be used not only to enhance the implementation of insights from the stock selection process and to control portfolio risk, but also to expand the range of risk-return tradeoffs available from the portfolio construction process. With short sales, it is possible to construct *market neutral portfolios* that balance the market value and overall market sensitivity of long positions against the market value and market sensitivity of short positions. The balanced long and short positions neutralize the portfolio's exposure to the underlying market, so the portfolio incurs no systematic market risk and earns no market return.

A market neutral portfolio can be *equitized* by purchasing stock index futures. The equitized portfolio will reflect the equity market's performance in addition to the performance of the long-short portfolio. As we discuss in "Alpha Transport with Derivatives" (*Journal of Portfolio Management*, 1999), the long-short portfolio's return from security selection can be "transported" to virtually any asset class that has viable derivatives.

Integrated Long-Short Optimization

We recognized early on that simply combining a portfolio of short positions with a separately optimized portfolio of long positions would not create an optimal long-short portfolio. Only if all potential positions were considered together, in a single *integrated optimization*, would the risk-reducing and return-enhancing benefits of short selling be maximized. We built a long-short optimizer that could integrate proposed long and short positions to take into account cross-hedging of positions.

Along with our work on complexity and on disentangling return-predictor relationships, our insights on integrated optimization are some of the most important work we have done. In "On the Optimality of Long-Short Strategies" (*Financial Analysts Journal*, 1998), we showed that long-short portfolios with any given exposure to the underlying market benchmark should be constructed with an integrated optimization that considers simultaneously both long and short positions and the benchmark asset. Rather than combining a long-only portfolio with a market neutral portfolio, it is better to blend active long and short positions so as to obtain a desired benchmark exposure. That article laid the foundation for the development of 120–20 and other *enhanced active equity strategies*, deriving precise formulas for optimally equitizing an active long-short portfolio when exposure to a benchmark is desired. "Long-Short Portfolio Management: An Integrated Approach" (*Journal of Portfolio Management*, 1999) provides another look at integrated long-short portfolios.

Once new prime brokerage structures were available to facilitate these strategies, we began to manage these types of portfolios, taking advantage of our insights into integrated long-short optimization. An enhanced active equity portfolio includes long and short positions and maintains a full exposure to an underlying market benchmark. In an enhanced active 120–20 portfolio, for example, an amount equal to 20 percent of the portfolio's capital is sold short, with the proceeds from the short sales plus the initial capital being invested long. The portfolio thus provides 100 percent net exposure to the equity market, along with many of the benefits that short selling allows in the pursuit of return and the control of risk.

In "Enhanced Active Equity Strategies: Relaxing the Long-Only Constraint in the Pursuit of Active Return" (*Journal of Portfolio Management*, 2006), we discussed these strategies and compared them with long-only and other long-short strategies. This article also highlighted 200–100 enhanced active portfolios. In contrast with equitized long-short strategies, which achieve market exposure with passive overlays of stock index futures or exchange-traded funds (ETFs), these 200–100 strategies hold active positions in selected individual equities.

We took a closer look at the relationship between enhanced active 200–100 portfolios and equitized long-short portfolios in "Enhanced Active Equity Portfolios Are Trim Equitized Long-Short Portfolios" (*Journal of Portfolio Management*, 2007) and demonstrated that an enhanced active portfolio is equivalent to an equitized long-short portfolio, with the two having the same active security weights and returns. The enhanced portfolio has the advantage, however, of being more compact and requiring less leverage. In "20 Myths About Enhanced Active 120–20 Strategies" (*Financial Analysts Journal*, 2007), we shed some light on this and other characteristics of enhanced active equity strategies that are frequently misunderstood by investors, including how the strategies increase investors' flexibility both to underweight and overweight securities and the potential benefits of using short selling and leverage to improve the risk-return trade-off.

Books and an Ethical Debate

Bruce: Back in 1999, I finally saw the fruition of a project I had been working on for years—the publication of my book *Capital Ideas and Market Realities: Option Replication, Investor Behavior, and Stock Market Crashes*. The seeds of this work had been planted in the 1980s, during some heated debates and discussions I'd had with Hayne Leland, John O'Brien, and Mark Rubinstein of Leland O'Brien Rubinstein Associates. (Rubinstein and Leland were also professors at the University of California, Berkeley.) They had devised a dynamic hedging product based on the Black-Scholes-Merton option pricing formula. When I had first joined Prudential Insurance, I had been asked to analyze this *portfolio insurance* strategy. I warned then that the strategy, although workable in theory, contained its own self-destruct mechanism. The strategy's automatic, trend-following trading could destabilize markets, causing the synthetic insurance to fail. Prudential had followed my advice, and even though in the short term they missed out on the management fees associated with a burgeoning portfolio insurance industry, they avoided the embarrassment and difficult client discussions after the strategy failed—when it was needed most—during the 1987 crash.

My insight was later recognized in *Pensions & Investments* by Editorial Director Michael Clowes, who noted that I was "one of the first to warn that portfolio insurance . . . probably would be destabilizing" ("More to say about crash," July 12, 1999), and in the *Wall Street Journal*, where Roger Lowenstein ("Why Stock Options Are Really Dynamite," November 6, 1997) said that I had "predicted before the 1987 crash that portfolio insurance would trigger chain-reaction selling."

After the crash, I saw the same dynamics behind portfolio insurance roiling the markets over and over again in other guises, including synthetic put options and the relative value arbitrage strategies pursued by hedge funds such as Long-Term Capital Management (LTCM). Prior to the collapse of LTCM, I had expressed my concerns in two *Pensions & Investments* pieces (Barry Burr, "Nobel-Winning Strategy Criticized," December 8, 1997, and Bruce Jacobs, "Option Replication and the Market's Fragility," June 15, 1998), taking issue with Nobel laureates Merton Miller and Myron Scholes (an LTCM partner). I last debated Rubinstein on the subject as a participant in *Derivatives Strategy*'s "2000 Hall of Fame Roundtable: Portfolio Insurance Revisited."

Apparently, many still fail to realize the limits of risk reduction and the potential effects of risk-shifting on market fragility. Systematic risk can be shared (with diversification) and it can be shifted (with options), but it cannot be eliminated. When too many investors forget this, risk in the market tends to build up, sometimes with explosive results. I discuss this problem in "Risk Avoidance and Market Fragility" (*Financial Analysts Journal*, January/February 2004).

In "A Tale of Two Hedge Funds," Ken and I describe in detail how the supposedly low-risk strategies of LTCM and another infamous hedge fund, Granite, came apart in spectacular fashion when they had exhausted the market's liquidity. "A Tale of Two Hedge Funds" appears in our edited volume, *Market Neutral Strategies* (2005), which brought together some of the industry's most successful practitioners to discuss long-short equity strategies, convertible bond hedging, and merger arbitrage, as well as sovereign fixed income and mortgage arbitrage. It serves as a cautionary reminder of how such strategies, when not managed carefully, can blow up, threatening the very markets in which they operate.

I had taken the liberty of sending a draft of *Capital Ideas and Market Realities* to Nobel laureate Harry Markowitz, who not only liked the work, but offered to write the foreword to the book. In Harry's subtle and piercing way, the foreword makes the distinction between portfolio insurance and portfolio theory and their effects on financial markets. Harry also wrote the foreword to a collection of Ken's and my most important articles, *Equity Management: Quantitative Analysis for Stock Selection* (2000) (also available in Chinese translation from China Machine Press). There he notes that our optimization work builds on his mean-variance theory and that some of his later work builds on what he calls our "seminal work" on disentangled expected return estimation procedures. In fact, we were surprised to learn that he had used our disentangling approach in researching and managing Japanese equities for Daiwa Securities Trust Company.

Harry's foreword to *Equity Management* echoes certain themes found in his foreword to *Capital Ideas and Market Realities*, in particular how the translation of investment ideas into products and strategies must involve trade-offs between theory and practice. Harry discusses why investors might want to add constraints on position sizes and sectors to the portfolio optimization solution, despite the theoretical cost of these constraints. Harry notes that our work on integrated portfolios and the estimation of security expected returns is "to be acknowledged for bridging the gap between theory and practice."

In the early years of the new millennium, it became apparent that the translation of theory into practice was fraught with other kinds of difficulties. In particular, the bursting of the Internet bubble revealed that Wall Street research, which in theory was objective and undertaken to benefit client portfolios, was in practice often conducted for the direct benefit of analysts and their employers rather than their clients. The CFA Institute proposed conflict-of-interest standards for security analysts' research and solicited comments from Institute members.

My response (August 12, 2002) noted: "Just as the research conducted by analysts at brokerage firms and investment banks is susceptible to the influence of commercial interests that may conflict with the best interests of their clients, so too the work done for and by [the CFA Institute] and its professional publications and conferences is susceptible to being influenced by interests that may conflict with the best interests of members and investors in general. I believe that the Research Objectivity Standards as proposed should be expanded to deal with these conflicts of interests." My proposal received substantial industry support, and the January/February 2003 issue of the CFA Institute's premier research publication, *Financial Analysts Journal*, announced new conflict-of-interest policies.

Portfolio Optimization and Market Simulation with Shorting

Bruce and Ken: After the publications of *Capital Ideas and Market Realities* and *Equity Management,* we collaborated with Harry on two projects of mutual interest. First, we investigated a tricky problem affecting the optimization of long-short portfolios. The optimization problem in general is tractable because certain shortcuts can be taken. Some models in wide use for long-only portfolios—for example, factor and scenario models—allow the investor to apply fast algorithms that greatly simplify the optimization problem. It is not readily apparent, however, that such models are applicable when portfolios hold short as well as long positions.

"Portfolio Optimization with Factors, Scenarios, and Realistic Short Positions" (*Operations Research*, 2005) and the less technical "Trimability and Fast Optimization of Long-Short Portfolios" (*Financial Analysts Journal*, 2006), which we coauthored with Harry, show that the same algorithms used for optimizing long-only portfolios can be used for portfolios that contain short positions—provided a certain condition holds. This condition, which we term *trimability*, often holds in practice.

We also began a longer-term project with Harry. This one also had roots in an area of past interest—the Black-Scholes-Merton option-pricing model that formed the basis of portfolio insurance. The model allows for the solution of option prices by assuming that underlying security prices change randomly and continuously over time. Such continuous-time models are useful because they can often be solved analytically. They are not useful, however, when investment actions or changes in the underlying environment alter the price process. Nor can they tell us whether microtheories about the behavior of investors can explain the observed macrophenomena of the market.

We developed a model of the overall market that has the potential to address these problems. The Jacobs-Levy-Markowitz Simulator, or JLM Sim, allows users to model financial markets, employing their own inputs about the numbers and types of investors, traders, and securities. The JLM Sim is an *asynchronous-time* simulation. It assumes that changes reflect events, which can unfold in an irregular fashion. Price changes may be discontinuous, gapping up or down in reaction to events.

The *Journal of Portfolio Management* article "Financial Market Simulation" (2004), coauthored with Harry, describes the JLM Sim. Those interested in finding out more about the simulator, or experimenting with it, can access JLM Sim at the Jacobs Levy Web site. We believe an asynchronous-time market simulator such as JLM Sim, which is capable of modeling the agents and market mechanisms behind observed prices, is much better than continuous-time models at representing the reality of markets.

Asynchronous models may also be superior when analyzing whether microtheories about investor behavior can explain market macrophenomena. From time to time, the market manifests *liquidity black holes,* which seem to defy rational investor behavior. One extreme case was the stock market crash on October 19, 1987. When prices fell precipitously and discontinuously on that day, rational value investors should have stepped in to pick up bargain stocks, but few did. Asynchronous models are able to explain both the abundance of sellers and the dearth of buyers.

Our experiments with the simulator show that only a relatively small proportion of momentum investors can destabilize markets, overwhelming value investors. Similarly explosive behavior can result when traders don't anchor their bid/offer prices to existing market prices. Using JLM Sim, we are currently examining the intriguing question of what conditions give rise to a stable equilibrium in the capital markets.

JLM Sim provides researchers with the means to create dynamic models of financial markets. It is our hope that their experiments will lead to more and more refinements in the JLM Sim, bringing its predictions into even closer alignment with observed investor and market behavior. In the long run, JLM Sim may become a powerful and reliable tool for testing the effects on security prices of real-world events such as changes in investment strategy or regulatory policy.

Looking back, the long days and weeks we have dedicated to our business have been more than adequately rewarded. We have produced 20 years of published research on our investment philosophy and, more importantly, 20 years of proprietary research for the benefit of our clients' portfolios. We have also given back to the investment community by supporting research that moves it forward. We were founding sponsors of the Research Foundation of CFA Institute and the Fischer Black Memorial Foundation. In 1998, we established the Bernstein Fabozzi/Jacobs Levy Awards for outstanding articles published annually in the *Journal of Portfolio Management*.

We feel fortunate that we can pursue our passion for equity research and portfolio management. We love being quants!

Chapter 20

Tanya Styblo Beder

Chairman, SBCC

I was born a world away from Wall Street, in Spokane Washington. My father was a medical doctor and my mother had been a high school teacher, educated in the classics and philosophy. When I was three years old my family moved to Saratoga, California—what would eventually become known as Silicon Valley—where I lived until I went to college.

Math ability must be in my genes, but that would not have been evident to anyone who knew my family. It was not just by professional choice that my parents were not mathematical. I remember, once, much later, trying to explain to my father the idea of higher orders of infinity, realizing after a lot of frustration that he was just not interested, and finally giving up the attempt. Both my parents were musical piano, cello and ballet were my main interests throughout my childhood. I remember, however, that while growing up I always did well in school in math. By default, I became the family math tutor.

I realize now that my thought processes work differently than those of the rest of my family. For example, when my sister and I were having trouble mastering a particular transition in a piece of music, my sister would work on hearing the transition, while I would work on visualizing the notes on the page. Today, I am likely to remember a person's phone number before I remember a name.

It was in my tenth grade geometry class where I first fell in love with the math. The teacher, Andy Dazols, was terrific and passionate. Somehow, he could make you really understand why parallel lines never intersected, and then, when he had made you a believer, he could make you understand why they all intersected.

Nevertheless, my math experiences did not lead me to math after high school. It was my intention, when I applied to universities, to study music and to eventually become a performance artist. Like most teenagers, I was anxious to attend a university far enough away from home that I would have to live at school. I applied to several West Coast schools, but my first choice was UCLA. I was accepted at Stanford but didn't want it for the simple reason that my nine-month older sister went there. When I was accepted at UCLA, my mother was opposed to my going there. I could not convince her to change her mind. It was getting late in the academic year, and I was desperate to get approval to go to UCLA.

I had recently come across an article in the *Wall Street Journal* that had said Yale was actively recruiting women from the West, as it had recently gone co-ed. Even though it was February, I applied, thinking naively that if I was accepted, I would have a bargaining chip with my mother, who surely would not want to see me go to the other side of the continent to attend school. I received an acceptance from Yale, and went to my mother with my strategic plan in full motion. Unfortunately, plans frequently do not seem to work out the way they are intended. Far from being horrified at the prospect of my attending Yale, my mother blessed the plan immediately, and I was soon saying goodbye to California.

Yale

Yale was a culture shock of major proportion. I knew nothing of the East Coast, the weather, the architecture, the people, the diversity of culture and opinion. I had gone to a relatively small and homogenous public high school, and from that pond I had been graduated as class valedictorian. But being "the brightest" was over—my freshman-year roommates at Yale were not only extremely bright and articulate, but it became quickly apparent to me that they were also better prepared academically than I was.

My class at Yale would eventually graduate ten theoretical math majors, of whom I and one of my freshman roommates would be the only two women. I didn't know that at the time, of course. I wasn't even planning to major in mathematics at Yale. I intended to pursue a career in performance music. Math was going to continue to be a hobby for me, because I enjoyed it and it was easy.

Another of my roommates also was intending to pursue music as a major, and she is at least partly responsible for my change in career plans. She was extremely talented—much more so than I. And I realized that no matter how hard I worked at it, I could not compete on the basis of raw talent. On a more practical level, music majors didn't have time for much fun, and that was a problem for me. Three or four hours of practice daily in a drab, windowless—and padded—room was draining and left little time for activities such as engaging in political discussions or just throwing a Frisbee on the quad. My voice class, a requirement for the music major, wasn't going well, either.

In contrast, my math classes went very well. Despite the disadvantage I had in my prior math training, I was able to do well without having to put in many hours of work. The choice was before me: Long hours of effort followed by a career where the chances of success and happiness seemed limited, or Frisbee on the quad and a happy and satisfying academic career in mathematics.

But it was also more than that. My math professors at Yale were inspiring. I was fortunate to have Serge Lang as the professor for my first math class. (At Yale, lower-level classes were regularly taught by the most senior professors.) Lang, who died in 2005, was known as much for his pedagogy as for his contributions to number theory. He taught me not to be casual in my mathematical thinking, even if the concepts came easily to me.

For example, I was asked to demonstrate a proof in class one day early in the semester. I went to the blackboard and wrote what I thought was a satisfactory proof. After I returned to my seat, Lang threw an eraser, and told me to go back to the blackboard and redo the proof, saying that it didn't matter that I had the right answer. It was as important that I methodically demonstrate how I got there. This was a pivotal lesson, because if I was to do research in mathematics, I needed to learn to be careful about and therefore confident in my proofs. It's not a bad life lesson, either.

My professors were also very supportive. Angus MacIntyre, the logician and algebraist, in particular, mentored me. In fact, my interest in mathematics was almost universally encouraged. With only a single exception I experienced no discrimination related to my being a woman. Even back in high school, my geometry teacher encouraged my interest by providing extra material (and homework) for me.

I took two game theory courses as electives, partly to avoid the normal senior electives like art history and physics for poets. Yale is famous for such courses, and they attract hundreds of students. My game theory courses had six. These courses were taught by staff of the Cowles Foundation for Economic Research. Martin Shubik, who was my professor, was at the time also the director of the Cowles Foundation.

As a result of taking these courses, I became interested in game theory and operations research. I had never taken the basic courses in probability and statistics, which were taught in the applied math department, and found myself again having to catch up to the other students. At Yale, theoretical and applied mathematics were housed in different buildings, and applied math seemed to be looked down upon by the theorists. It was Shubik who told me, as I was struggling to catch up, that it wasn't more math that I needed to learn—rather, I really needed to learn more about how to use mathematicians.

I had never doubted that I would go on to study mathematics in graduate school. But my growing interest in game theory, combined with the departure from Yale of some of my math department mentors, especially Angus MacIntyre's moved to Oxford, put me in a frame of mind receptive to a different focus of that future graduate work. Further, I felt that game theory offered more open questions than mathematical logic, which had been until then my intended future field of study.

Martin Shubik guided me in my applications to graduate school. He helped set up interviews for me with faculty at several other schools.

However, in the end, perhaps not surprisingly, I decided to stay on at Yale for my graduate work, and to work with Shubik on my dissertation. But Fate was about to interfere with my plans once more.

The Yale administration announced that it was creating the School of Management, and it was intended that the Operations Research Department would become part of the school. At the news, there was a significant '"wait and see" approach among the Operations Research faculty. It was not at all clear whether there would be a mass exodus of faculty to other schools. Shubik recommended that I hold off for a year before starting graduate school in order to allow the situation at Yale to clarify itself.

This presented me with a problem. It was March of my senior year. Employment recruiting among the undergrads had long since been completed. When I went to the career office, I was told there was only one company that was still taking student resumes, a company named First Boston.

I had no choice. I applied, not only not knowing what the job was, but also not knowing that the job was located in New York City and not in Boston. After an initial interview on campus, I was invited to New York for additional interviews.

When I returned to New Haven, I had learned where the job was located, but I still didn't know anything about the job. I had interviewed with several managing directors, but the conversation was mostly about books, sailing, and travel. An offer letter soon followed, and I became one of four analysts hired by First Boston that year.

First Boston

I was assigned to work with Joe Perella in the fledgling M&A Department. Joe himself had joined First Boston as an associate a few years prior, in 1972. Joe eventually would become one of the biggest names in M&A, but when I joined his group, it numbered just a dozen members. I had no idea what M&A was or how it worked.

I was put to work analyzing financial statements. The mathematics required for this activity was decidedly limited, but I found it fascinating. The amount that you can learn about a company simply through manipulation of its financials is amazing. Of course, before I could do any analysis, I had to go to accounting boot camp. I enjoyed learning accounting, because it is so well-ordered, so rules-based. I put that new knowledge to good use on one of my first assignments, the Kennecott Copper-Carborundum merger.

Perella had an amazing feel for a company's financials. He would look through a set of financial statements and estimate the company value within a few percent of the number that I would arrive at after a considerable amount of work (this was before the time of desktop computers).

Working in M&A was a heady experience. The deals were big, complex, and very newsworthy. The hours were long and there was much for me to learn. Sometimes the learning was humbling. About six months after I started, I was scheduled to participate in my first board presentation on a proposed merger, along with Perella. As we were in a car on the way to the meeting, Joe was reviewing the presentation for the first time. After a little while he looked up and asked me how I had gotten these numbers. After I explained, he told me that what I had done was wrong. As Joe began the presentation to the board he said that the numbers in the presentation were incorrect, and that Ms. Beder would explain why and what the correct numbers were. He had not told me beforehand that he was going to do this. The board members were very kind to me that day.

A year passed, and I didn't head back to Yale. Instead, I headed to London to work on cross-border M&A deals, and I continued doing this for two more years. During this time, the M&A group grew to 40 people. More importantly, the character of the business changed. When I had arrived, the central merger question was "how much is the company worth," but that question had been replaced with "how much will it cost to buy." People attached value to the size of the news headlines that deals generated. I was looking to move out of M&A, but nothing was available at First Boston, so I looked to New Haven again.

Graduate School

I returned to Yale, and left First Boston on a leave of absence. I arrived in New Haven with an intended topic for my thesis, specifically, a game-theoretic approach to the valuation of control blocks in takeovers, which I would write under Martin Shubik.

Before working on the thesis, I had to get beyond the first year of graduate mathematics classes. After three years away from math, it really was like Greek to me. At the end of that year, I presented my thesis proposal to the PhD program thesis committee, only to have it rejected. I was told that the topic was too applied.

With Shubik's help, I transferred to the business school at Harvard and entered its DBA program. Martin and I have kept in contact over the years, and he to this day encourages me to return to academia.

The environment at Harvard was very different. To the other students, I was one of the two weird math geeks.

Every DBA needed first to complete at least the first year of the MBA, and that's how I spent the next year. I loved the financial analyst/financial analysis coursework plus the competitive analysis courses. But I had a hard time embracing organizational behavior. I was one of the profs' favorites to call on in class because I always felt the right action was to fire the "problem" employee. Reflecting on that now, I wish that in my career I had been quicker to gain an appreciation of the messages of that course.

Like most students, I was short of money, so in the summer after my first year at Harvard I worked for McKinsey in management consulting. But by the time the summer ended, I was back at First Boston. When it was time to return to Harvard for my second year, it was with an agreement with First Boston to continue to work for them part time during the school year.

During my second year at Harvard, I did a specially directed studies course under Michael Porter and worked on some fascinating strategic vision projects regarding data, technology, and competitive practices. But the lure of Wall Street was very strong.

My time at First Boston now was largely taken up by a project to study interest rate swaps. Swaps were a new product, traded by appointment, with the business being done "upstairs" in the investment bank, not on the trading floor. When I finished my MBA, First Boston asked me to join full time to put together swaps deals in the Corporate Finance group. I agreed, and there ended my pursuit of a PhD in mathematics.

Six months after rejoining First Boston, I received a phone call while on the road informing me that I had a new boss and a new job. I had been transferred to the trading floor to partner with an experienced trader on the new swaps desk. After the initial shock of the change, I found that I loved it on the trading floor.

The trading floor was a different world in many ways. It was not a politically correct environment. The other traders liked to tease me. I stood out with my navy blue suits and ridiculous silk bow ties that women wore in the 1980s. The trading action was slower, with much less volume. A $10 million swap trade was a big event. The trading perspective was also different from corporate finance. I learned from my own experience—including my mistakes which as a trader involved my own P&L.

Swaps

When I moved to the floor, swaps trading was in its infancy. Technology and analytical tools available to us were extremely primitive. For example, I was on the trading floor when the world got zero coupon yield curves, when we left Monroe bond calculators behind, when volatility got more than the two states of 1 and 0, when we got spreads rather than absolute prices, and when callable bonds got option-adjusted spreads rather than trading to the call or maturity date, depending on what was in the money.

In 1984 I wrote an article in the *Harvard Business Review* called "How to do Swaps," one of the very first articles written on the topic. This was published only after fighting with the editor for several months over the closing paragraph that predicted that swaps trading would be a $25 billion market before too long. A few years later the street would do $25 billion for a new derivative by lunch.

Over time, other products were added to my responsibilities, first currency swaps then caps, collars, and floors. Later, I moved to fixed income research and spent time on applications of derivatives in general (then called synthetics) to hedging mortgage-based instruments and as well as asset/liability problems.

I really liked using derivatives to help to solve asset/liability and other problems. I decided to leave First Boston at the end of 1986 to start a one-person risk-management consulting firm. My timing was so perfect that if I were a less humble person, I would claim prescience as one of my skills. When the stock market crash happened in September 1987, everyone started to focus on risk management and derivatives, and it was a very exciting time.

My first consulting job was for a major insurer. I was asked to figure out how big the embedded options were in certain guaranteed investment contracts, and then how to hedge them. My second job was designing the asset/liability management program for a savings and loan. Other projects followed quickly: the design of VaR for several Wall Street houses, portfolio challenges caused by kitchen-sink bonds and inverse floaters, Asian currency plus derivatives portfolios, correlation challenges within global businesses. I liked working on new problems with each engagement, and in my 13 years of consulting I was pretty much able to do that, and generally each had a significant quant component.

Giving Back

Teaching has always been part of my professional life. While I was still at First Boston, the head of equity research asked me to give a guest lecture in his course at Columbia. I loved it. I taught a course of my own at Columbia in the mid 1980s and then started teaching as an adjunct at Yale. At Yale I had the pleasure over 15 years of coteaching with both my early mentor, Martin Shubik, and later my former Yale classmate but now well-known professor Will Goetzmann. In 2005, I returned to teach at Columbia at the encouragement of Emanuel Derman, along with Leon Metzger.

In the early 1990s Jack Marshall and Bob Schwartz asked me to become involved in a professional society they were then organizing, the International Association of Financial Engineers. I was enthusiastic about being a part of that organization as I love everything about financial engineering. Eventually, I served as Chair of the Executive Board, and I am still very much involved with their work as a Director and co-Chair of the Investor Risk Committee.

I left consulting in 1999 to return to a trading and business building role—this time in the hedge fund industry for Caxton Associates.

Five years later, I moved on to organize and build a boutique but institutional-quality multistrategy hedge fund inside a large financial institution, serving as the CEO of that business. Quantitative finance figured prominently in both hedge funds. The best part of my career has been to do what I love and have a passion for. It's a daily joy to watch the plan you wrapped around a vision become a reality. To be so lucky to work with a terrific team that you get to pick is priceless, I am blessed to have had the opportunity to share my days with such amazing colleagues. I love the math, but it's really about the people.

Today I am planning the start of the fourth big business build of my career, again including a good dose of quantitative finance. I continue to teach and research, and I am pleased to give back to the industry through my roles with the International Association of Financial Engineers and the National Board of Mathematics.

Chapter 21

Allan Malz

Head of Risk Management, Clinton Group

I wasn't raised to be a quant. My childhood milieu recognized two professions as legitimate: doctors and lawyers. In spite of an early interest in science, and an aptitude for well-formulated expressions of antagonism, I wasn't attracted to either one.

Mathematics took a while to tighten its grip on me. I became interested in economics in high school to the increasing the puzzlement of my family. In spite of the reputation of economics as a somewhat quantitative discipline, I had little interest in mathematics. I did well in algebra and geometry, but to this day I have some problems understanding trigonometry, which has always seemed somewhat mysterious to me. My indifference to mathematics continued through college. I had the additional good fortune to attend a university that had abolished all quantitative requirements for an economics major.

How Not to Get a PhD

Graduate school broke in on my mathematical slumber like a death knell. In the first of three attacks, the last successful, upon the task of gaining a doctorate, I enrolled at Princeton. This institution was apparently as unaware as I was of my lack of preparation for graduate work in economics. My first inkling of the gravity of the situation came on my first day of class, in Microeconomic Theory. The professor drew the letter A on the blackboard and described it as a *matrix*. I had vaguely heard of matrices, which I thought had something to do with either biology or sociology, but never as a mathematical concept. The letter surprised me; I would have expected a picture of a box or a beehive.

Rudely awakened, I began a speed course in math. Alpha Chiang rescued me from what seemed a certain fate. I learned most of the math I know today in the next two months. In spite of the circumstances, I even found the experience enjoyable. Like Winston Smith via Room 101, I learned to love mathematics. I completed the academic year honorably, left the university, and spent the next decade wandering around the Middle East and Europe.

In the course of my wanderings, I sojourned at Ludwig-Maximilians-Universität in Munich, less well-known as a center of economic research than as the place where Sophie Scholl distributed anti-Nazi leaflets on a Thursday in 1943, leading to her execution the following Monday. At the time, West Germany provided a free education, residency, and work permission to qualified foreign students. I doubt I fit the profile of the intended beneficiaries.

My romance with the mathematical sciences continued. To finance my studies, I worked with a professor attempting to demonstrate the existence of business cycles of fixed periodicity, a theory that had the benefit for me of requiring an almost infinite supply of research assistant and computer time. Economic science may have benefited less.

Computers were becoming rather a big thing at this time. My first encounter with them, at Princeton, had been via punch cards, which I still remember fondly. They ordained a certain parsimony and elegance in programming, like having to write with a very thick pencil.

It was swell to be an American in Europe during the Reagan era. My fellow students saw the United States as a fascist state likely to turn their continent into a sheet of radioactive glass. I respected their views, since they seemed to be more conversant with American popular culture than me.

How Not to Get a PhD, Continued

After harvesting a degree from Munich University, the time came to return to the States and see if I could earn a living. Although it had nothing to do with my personal choices, The Great Moderation had begun, and the prospects for finding work were far better than during the leaden 1970s. Back in New York, my first job was as an economist at the Federal Reserve Bank of New York, in a department quaintly named the Industrial Economies Division, in contrast to the Developing Economies Division. Today, I suppose, I would be working in the Service Economies Division and the emerging markets area would be Industrial Economies Division. The group's responsibility was to forecast GDP growth, inflation, and macroeconomic conditions generally in Western Europe.

At the same time, I enrolled, again, in a doctoral program in economics, this time at Columbia University. Even against my history of poor choices, the decision to simultaneously begin a PhD program and a professional career stands out. I didn't receive a degree for another decade. I took as my dissertation topic monetary issues in Europe, where several industrial economies still remained.

Years passed, and the fascination of watching the dollar and the price of oil fall began to fade. I switched to a different department at the New York Fed, this one tasked with studying financial innovation. It had its origin in a Bank for International Settlements group that also had a quaint name, the Eurodollar Standing Committee. The Committee dated back to concerns about one of the first great financial innovations of the postwar financial system, the Eurodollar market.

Today the Eurodollar market would hardly evoke a yawn from most observers, but at the time the central bankers meeting in Basel took an interest, it was considered a great mystery and even a threat. The Eurodollar market had arisen as a side effect of the U.S. current account deficits that had brought down the original Bretton Woods system, it appeared to threaten a loss of the Federal Reserve's ability to limit the U.S. money supply, and to top things off, the Soviet Union and later the oil exporters were among the largest depositors.

In my new role, I searched for reasons to worry about financial innovations that, like the Eurodollar market in its time, were poorly understood, and that with some imagination could be seen as potentially harmful. In particular, I studied newfangled currency options such as average-rate, basket and barrier options. And, like the Eurodollar market, these now seem about as threatening to financial stability as a plate of pasta.

This was my first encounter with modern finance. The theoretical and institutional basis for it had been established over the previous two decades, and it was just beginning to emerge as a really large field. I found it attractive because it is pretty and has a lot to say, though admittedly about a very small field of human activity.

I also made my first acquaintance with a new discipline within finance, risk management, through which banks involved with new financial instruments were trying to cope with difficult-to-understand exposures. My boss at the time was excited by a paper he'd received from First Chicago, outlining an approach to measuring the volatility of a derivatives portfolio using the variance–covariance matrix of the positions' underlying returns. Value at risk seems to be one of those ideas simultaneously and independently born in a dozen brains. I personally know several people who discovered it.

I continued to meet my new objectives of acquainting myself with modern finance and my old one of procrastinating on my dissertation. A new opportunity for delay arose in the New York Fed's foreign exchange department, which had decided that economists might make good foreign exchange traders.

The New York Fed's FX desk is responsible for carrying out currency intervention operations on behalf of the U.S. Treasury and the Federal Reserve System; the Treasury owns the policy, the Fed the machinery. Most of the Fed's FX transactions are essentially commercial, executed on behalf of central banks that deposit reserves with it. By volume, however, the sporadic but large policy-driven FX market interventions predominate.

I had the good fortune of arriving on the desk in mid-1992, in time for a sustained bout of intervention in support of the dollar that lasted until early 1995—in fact, the last such sustained sequence to date. Intervention appeared to occur primarily to allay policy makers' fears that markets would get the impression that nobody's home, rather than to reach or defend a particular exchange rate level.

Participating in the FX market brought me back to the perennial economists' debate over market efficiency. My observation was that they are, but that there is a messy and hard-to-understand process for percolating information among participants. Sooner or later, market participants changing positions and investing a lot of resources get the market close to the right price, but no price is right for long. What is different about a day on which buying $100 million doesn't prevent the dollar from falling 2 percent and a day on which buying $50 million drives the dollar up 3 percent?

An episode I found instructive took place in the summer of 1993. The yen had been accelerating a steady rise against the dollar, and the Treasury did not want the exchange rate to go below 100. As it happened, the entire Fed trading room was either on vacation or otherwise engaged, apart from myself and another trader. We received a fairly large authorization to buy dollars. On the first round, the dollar rose much more sharply than we anticipated, and the markets were convinced that we had bought a large multiple of the actual intervention. We stopped buying, but the dollar kept going, and rose 4 percent (about six to eight standard deviations). Needless to say, on other days, an intervention many times larger had no impact on the market.

I grew increasingly interested in using market prices to draw inferences about what the market is thinking about future prices and the balance of risks. Our FX counterparties sent us reams of faxes each day reporting on the markets, including runs of currency option prices. One option combination, the *risk reversal,* by which counterparties trade an out-of-the-money call for an equally out-of-the-money put, was particularly intriguing, especially since no one in the trading room had a clue what it was. Once I understood what this beast was, I realized that it had a lot of information about the probability distribution of future exchange rates the market carried in its collective head. It even lined up well with odd outcomes of currency market interventions. I spread the word to Fed and foreign central bank colleagues, one of whom charmingly and appropriately referred to this option combination as a *role reversal*.

In a moment of rapture, I also realized that I could write my dissertation on this. My topic, after all, was on forecasting realignments in the European Monetary System. I had flailed around for years with various regressions and Kalman filters. My advisor, who had previously shown no great enthusiasm for my infrequent visits, heard me out on the subject of risk reversals and told me I'd be done as soon as I wrote it up. And so it was. I got my degree a few months later.

RiskMetrics' Salad Days

By the time I took this step toward certified quanthood, I had been at the Fed for close to a dozen years. I started to make discreet inquiries regarding alternative employment possibilities by telephoning every finance and economics professional I had ever even fleetingly met. I was particularly interested in risk management.

I eventually landed at Credit Suisse First Boston for an interlude as a fixed income risk manager and for a bizarre experience of management at its worst. The head of risk management thought I was quite a find, but hadn't thought to ask my future immediate supervisor if he agreed. Nor did he think it necessary to have me meet in advance with my own direct reports. The range of his outside activities also impressed me, and I understand he later left the firm under a cloud.

I did, however, learn one vitally important lesson: It is far more difficult to create a system that delivers even simple risk information than it is to invent interesting statistics. In fact, most of my time was spent chasing down files from dozens of First Boston systems, some of them dating back to the colonial era, all of which had to enter the grand central processor before one could reliably say anything about the portfolio.

Fortunately, a group at J.P. Morgan dusted off my resume, which had continued to molder on various desks around Manhattan. The group was responsible for consulting on risk management to Morgan's external clients. It was also the custodian of an Excel-based value-at-risk calculator called Four-Fifteen, named after the time of day by which risk reports ought rightly to be delivered, and of a production process for variance-covariance estimates for several hundred market risk factors, updated daily. In September 1998, it spun off as an independent, privately owned company, with me among the initial 25 members.

The spinoff took place close to the height of the first Internet boom, a time at which companies that had a single ridiculous idea could trade publicly at equally ridiculous valuations. RiskMetrics Group had many ideas, and only a few of them were ridiculous. Initially, we were harbored in Morgan's Wall Street headquarters, and our startup tech company atmosphere stood in marked contrast to staid Mother Morgan.

We were not the first company to specialize in risk measurement and reporting software, but the existing systems were heavy and costly to install and maintain, and were therefore commercially vulnerable now that delivery of information via the Internet was becoming feasible. RiskMetrics grew steadily, to over 200 people and many millions in revenue at the time I left, but it never quite lost the campus feel of a tech company.

Changes in the regulatory environment were also helpful to RiskMetrics' business. The Golden Age of Bank Supervision was arriving, when it seemed as though a stable relationship would be established between regulators and financial institutions. The regulators would finally get their arms around what those traders were up to, but the markets would be free to innovate. The regulators would use subtle, indirect tools such as capital requirements, rather than the blunt tool of prohibition.

The new approach started in the 1980s, as a reaction to a series of bank crises, focused in different phases on a range of bank assets, from real estate loans to emerging-market debt. By the time these crises had passed, banks had turned themselves from lenders into packagers of loans, investors, and providers of services. Regulators shifted the focus of regulation from static rules and solvency tests to capital adequacy, and eventually, to capital tests based on risk measurement. If we're going to protect banks with deposit insurance and an implicit too-big-to-fail guarantee, they said, let the banks run themselves so as to minimize the likelihood that their equity disappears and the taxpayer steps in. From RiskMetrics' point of view, it was easier to sell risk analytics to satisfy a regulatory requirement than purely on the strength of the information.

No More Mr. Nice Guy

My most recent shift has been to the hedge fund industry, as the risk manager for the Clinton Group. After the sedate and thoughtful atmosphere of an analytics and research company, a hedge fund offers a refreshing cold plunge.

Hedge funds are the most recent financial innovation to be digested by the broader public. They are nonetheless a fairly old institution—in fact, almost as old as the Investment Company Act, which didn't anticipate them. I suppose that makes them the world's second-oldest form of pooled investment. But they have become an important financial services sector only in the past two decades or so. Among the main drivers of their growth have been the development of derivatives and structured products and the finance theory that supports them, since these products provide the perfect playing field for hedge funds, and mutual funds aren't well situated or permitted to get involved. The other main driver of growth has been the fact that the world is becoming both richer and less volatile, driving down risk premiums and returns on conventional assets and making it harder for individuals and pension plans to profit: it used to be much easier to be a rentier.

Hedge funds have always practiced risk management, in the sense that they examine their portfolios carefully to make sure that the risk/reward relationship is the one they want, and that they don't blow up. But as long as the money belonged to the manager and a few other people, and the fund was not that large, this could be done by inspection. As hedge funds have grown, and as they have begun to enter institutional portfolios, the need for a specialized risk management function has grown.

The role of risk manager in a hedge find has a number of benefits. For one thing, all of the traders hate you. Another is the fact that, in the nature of random things, the worst loss is always in the future. Best of all, I can experience first-hand the limited usefulness of quantitative information.

On this sobering topic, I've drawn two main conclusions. First, without quantitative information of some sort about the portfolio, a risk manager can't know or do anything. Yet almost all the information one generates is deeply flawed, and is more likely to be flawed, the more sophisticated it is. Second, not only is it inevitable that the information is deeply flawed, but it is also not such a bad thing.

I've been fortunate to be in at least some of the right places at the right time: at the Fed during the Golden Age of Supervision, at RiskMetrics during and after the technology bubble, and at a hedge fund during what may prove the heyday of hedge funds. Perhaps it's the distorted perspective that comes from studying modern finance too much, but it seems to me that my career has been driven a lot more by my response to random events and larger forces than by my personal attempts to shape a path. Almost like life.

Chapter 22

Peter Muller

Senior Advisor, Morgan Stanley

I've had a fun and lucrative career as a quantitative trader. I built and oversee a group at Morgan Stanley called *Process Driven Trading* that trades firm capital using quantitative models. We have a culture that tries to merge the best of academia and Wall Street. Our track record is not public, but it compares favorably to the best hedge funds.

My journey there involves a few small random events that changed the course of my life. I'm not sure the story will help others who want to do the same thing, but hopefully it will be somewhat entertaining.

What's that Smell?

I hate cigarettes. I mean, I really hate cigarettes. If I'm in a restaurant and someone lights up across the room, I'll know, and it will make my meal less enjoyable. When U.S. airlines finally banned smoking on all flights, I threw a party. I hate cigarettes.

It was a cigarette butt in an ashtray at the bathroom of BARRA that almost caused me not to interview. And if I hadn't interviewed, my life would have taken a very different direction.

It was October 1985. I was living in California, playing piano for a rhythmic gymnastics team, slowly realizing that I wasn't making enough money to eat and pay the $200 a month rent that a friend of a friend was charging me. Then there was my friend's crazy roommate who would occasionally fire rounds off of a shotgun into the air when he was feeling depressed. I needed to get a real job.

After graduating with a math degree from Princeton earlier in the year, I had driven across the country to California. Once I got there, I had trouble getting my car to turn around and head back East. I had a good job offer waiting for me in New York from a German software firm, Nixdorf, where I had worked the previous two summers. But I didn't want to work in New York, not even for their very generous offer of $36,000 a year. I had fallen in love with Northern California.

So I kept pushing back my possible start date at Nixdorf and interviewing for programming jobs in the Bay Area. I even applied to IBM, whose corporate culture was not a match for me, but fortunately they told me that $36,000 was way too high a starting salary for them. I answered an ad for a small company in Berkeley that was looking for a Fortran programmer. I didn't know Fortran, but figured that since I knew a few other languages, picking it up wouldn't be a big deal.

Which brings me back to the bathroom at BARRA and the cigarette. While I was willing to work for a big corporation in order to stay in California, I was not going to work anywhere where it was okay to smoke. Somewhat reluctantly, I went through with the interview, which was fortunate, since I soon found out that smoking was not allowed at BARRA and that it must have been a visiting client sneaking a bathroom smoke.

I interviewed at BARRA with seven or eight smart and interesting people, about a quarter of the company at the time, and it soon became clear that this could be very fun job for me. Even though I had never taken a finance course at Princeton and was even a bit of a socialist (I remember making fun of my girlfriend at the time for taking a summer job at the *Wall Street Journal*), I was a math geek and theoretically intrigued by how money worked. I have an early memory of a family vacation in Europe where I asked my dad why exchange rates were different in different countries and didn't that mean there was a chance to make money by buying Deutschmarks in London and exchanging them back in Germany. I think I was 10 at the time and had never heard of the concept of arbitrage.

After my last interview, I realized I had been left alone at BARRA. I wandered the halls but it seemed everyone was out to lunch. All but the CFO, Ron Lanstein, who kindly took me along to lunch with him.

BARRA was using math to help analyze investment decisions for large institutional portfolio managers. The firm was started by Barr Rosenberg, a Buddhist finance professor from Berkeley and a very deep thinker. Barr's Buddhist beliefs shaped the culture at BARRA—everyone was treated equally (travel allowances, hotel accommodations, vacation policy) with the single (important!) exception of compensation. People were happy there, and this made it a fun place to work.

Barr had a brilliant knack for interpreting data. I didn't get much chance to work with him, but I do remember an early research project I did when he was still at the firm. It was an analysis of principal components of equity returns—I showed Barr the results and was impressed by his instant deep analysis of the data that made it come alive. ("This factor must be oil prices—look at the spike during the energy crisis, . . . and this one must be related to interest rates.") When I went back to do some more work the next day, I realized that I had made a rookie mistake and the data I had shown Barr was garbage. Oops! I fixed the analysis, and then, somewhat sheepishly, showed Barr the new time series. He didn't miss a beat "This makes much more sense. This factor must be the oil factor . . . and here's where the Fed came in and tightened . . ." (To be honest, I've told this story so many times I can't remember any more if it really happened this way. But I include it because it captures Barr's ability to make data come alive.)

Barr had sold BARRA to Ziff-Davis, a media conglomerate, in 1981, and convinced it to invest in building the next generation of equity risk models. After five years, Ziff was frustrated that the new models were not producing enough revenue and wanted to sell the company. Andrew Rudd, BARRA's president at the time, subtly let Ziff know that they would reject any inappropriate buyers, and after a number of potential buyers got scared off, the employees ended up buying back the firm for approximately the same price it had sold for five years previous.

I fell in love with the culture at BARRA—smart people who lived well and were working on interesting problems. Fortunately, they liked me, too. There was only one problem. As part of the sale back to employees, Ziff had prescribed a hiring freeze until the deal went through. So although they wanted to hire me, they couldn't. After waiting a couple months, I finally gave BARRA a fill or kill. Fortunately for me, they found a way to hire me on. And using my Nixdorf offer as leverage, I negotiated a starting salary of $33,000 a year. It seemed like a lot of money at the time.

Life at BARRA

During my first six months at BARRA, I quickly realized that I wanted to learn a lot more about finance. I had learned FORTRAN and had made some contributions fixing some code, but I realized that I wanted to be working on the most important stuff going on at the firm. My music interests took a back seat and I jumped into learning about portfolio theory.

I was very fortunate to find not one, but two mentors at BARRA. The first was Arjun Divecha, who headed BARRA's client support area. Arjun had a great ability to explain quantitative models in an easy-to-understand way. When he left BARRA, Arjun joined GMO and launched the most successful emerging markets fund ever. As of this writing, he runs approximately $15 billion and has an unmatched 10-year track record.

My second mentor was Richard Grinold, famous for his book with Ron Kahn on Active Portfolio Management. Richard heads all quantitative research at BGI, as of this writing the largest investment manager in the world. Richard and Arjun both wanted me to work for them. Andrew decided I was to work for Richard. He was a great boss. I learned a lot from him and have great memories of brainstorming sessions and running in the Berkeley hills at the end of the day.

Richard, with a background in operations research and time in the military, was great at practical solutions to hard problems. I asked a lot of questions doing research projects for him, and absorbed a lot of econometrics and portfolio theory in the process. As more of a pure mathematician, I sometimes had trouble grasping how Richard got from one equation to another. Our communication improved greatly when I got him to start using $\approx$ (for approximately equal to) instead of $=$.

After two years at BARRA, I asked Richard if he thought I should get a PhD in finance. I was one of his key employees, and the question might have put someone else in a tough position. Richard immediately told me that I should get the PhD for my long-term development and that I could always come back to BARRA. I remember being impressed by the unselfishness of the gesture. (Then again, maybe he was just sick of me.)

I decided to audit a course at Berkeley before committing to a PhD, and I'm very glad I did. After a few classes, I realized that we had better data than was available to the Berkeley grad students and that we were doing more interesting applied research into portfolio theory. I had also already read many of the papers they were discussing in order to do my work. The academics seemed very naïve about the real world. Funny how things are relative. Years later, when I joined Morgan Stanley, I realized how naïve we were at BARRA about how markets really worked.

So instead of getting a PhD, I worked even harder at learning quantitative finance. I started giving talks at BARRA's annual research conference and focusing on original research. One of my best papers was on biases in quadratic portfolio optimization. The basic idea was that since covariance matrices are estimated with error, minimizing portfolio variance would result in error maximization, and thus underprediction of portfolio risk. I was trying to figure out if other portfolio construction techniques without this bias did a better expost job of building minimum risk portfolios. Fortunately (since BARRA made money selling portfolio optimizers), the answer was no. (Years later, at Morgan Stanley, in work that will remain unpublished, I found a different answer.)

The BARRA culture was very casual—no one except for the head of marketing wore a suit and tie to work, and we would all go out to lunch for an hour and BS about world events or academic theory. There was a shower in the office, and late in the afternoon a bunch of us would go out for runs in the Berkeley hills. Once a month we would do this at night under the full moon and then go out for ice cream or beer.

BARRA was full of young professionals, and almost equally divided among the sexes. This, as you might imagine, led to a number of internal mergers. I managed to keep my own relationship secret enough that when we finally moved in together, people looked at the housewarming invitation and asked, "Is this Peter from BARRA or Bonnie from BARRA?" not realizing that two people they were working with had been dating for a year.

There may have been a few Republicans at BARRA, but they stayed in the closet. Our receptionist was named La Gerald. People wore costumes to the office for Halloween. La Gerald's involved a boa . . . let's just leave it at that.

One of the more unusual projects I worked on at BARRA was a consulting project we got from an obscure (at the time) futures trading hedge fund. The Renaissance Medallion fund did not have to use most of its capital for margin and wondered where on the yield curve it should invest the extra. I'm not sure I gave that brilliant an answer, but the company was impressed enough to offer me a job. I was tempted, since there were a bunch of smart mathematicians in the group, but I believed in efficient markets at the time, and I didn't think that math was going to help them beat the market. This was 1989 and efficient markets were almost a religion at BARRA, even though many of our clients were active portfolio managers.

You Gotta Know When to Fold 'Em

BARRA grew a lot in the seven years I was there, going from 30 to 200 employees and increasing revenues manyfold. I was lucky enough to be given a chance to buy some stock early on, and as a result, had a nice windfall when BARRA went public in 1991. But going public changed the firm. A bunch of senior people all of a sudden got liquid and started working less hard. The transition presented some difficult management challenges that were not adequately addressed. As a result, the people at my level started getting restless and a few left.

I also started working less. Along with a couple friends of mine at the firm, I picked up a new hobby—poker. The Oaks Card Room in Emeryville was open 24 hours a day, 7 days a week, and only 20 minutes from BARRA. I read a few books and started playing at small stakes. After achieving some success, I eventually moved up to the largest stakes available at the Oaks.

Poker developed into a bit of a habit. I was playing 10 to15 hours a week on average. On a couple of occasions, I remember going with my friend Claes at 6 P.M. Friday after work and playing until 10 A.M. Sunday. In a casual conversation with Richard, I suggested that if I gave up poker and put the extra 10 to 15 hours I was spending on poker into BARRA, I probably wouldn't make any more money. Richard agreed.

My last year at BARRA I earned $100,000.05. I remember having a conversation near bonus time with Andrew, where I indicated that Richard had told me I was going to make six figures that year. From his face, I can only guess at the subsequent events: A stern conversation with Richard ("Why did you promise him that?) and a recalculation of what my bonus would have to be. I guess they got it wrong and overpaid me by five cents.

I had a very nice life, a home in the Berkeley Hills, a great girlfriend, good friends at work. I had started playing music again and had a jazz group I jammed with on a weekly basis. And while I was ambitious, I was certainly making enough money.

But the fire had died at BARRA, and I started thinking that maybe it was time for me to leave, too. In searching for something new, I came up with a great idea—my poker buddies and I could take BARRA's quantitative approach to forecast returns as well as risk, and use it to trade capital. We were all pretty excited about it. Unfortunately, it was not a business that BARRA had tried, and we had just raised money in the capital markets without mentioning that we were about to start an internal hedge fund. I tried hard but couldn't sell the idea.

The alternative path given to me by Andrew was to develop models forecasting return and sell them to clients, but this wasn't nearly as interesting. If you could really forecast returns, why not use your forecasts to make money directly? And if you couldn't really forecast returns, you were selling snake oil (or mutual funds).

Despite these reservations, I launched the Alpha Builder group at BARRA. I hired a friend, Mark Engerman, who built it into a nice business after I left. Mark eventually went on to run money at Numeric Investors in Boston and then, in a move I have tremendous respect for, decided to quit his job and teach high school math.

The Call that Changed Everything

I was trying to figure out what to do next when I got a call from a Chicago headhunter looking to fill a quantitative strategist position at Morgan Stanley. It was a reasonably high-profile position, publishing research for the firm's clients, but I never thought I wanted to end up on Wall Street.

Still, I was searching, so I decided to fly to New York to interview. It was a pivotal decision for my life. At the time, although quants were a big part of fixed income departments, they hadn't yet infiltrated equity divisions on the street. Having an MBA had only recently become more important for sales and trading in equities—PhDs would not start showing up regularly on equity trading desks for 5 to 10 years.

It was brisk spring day in May 1992. I interviewed with a number of senior people in the derivatives part of the Equity Division. At the end of the day, I was asked to talk to someone who had a job title fairly similar to the one I was being offered. It didn't take too long to figure out that they were thinking of replacing him with me.

Not only did I not want to take his job away, but I really wasn't interested in the job for myself. It seemed they wanted a quant salesperson to engage clients in conversations to show them how smart the firm was. But ultimately, Morgan would make money trading for the clients and charging commissions. While coming up with money-making ideas would lead to better relationships and better business for the firm, it was a second order effect, and ultimately, not that different from something like Alpha Builder.

So when they asked me if I wanted the job, I told them no thanks. But the Morgan folks were persistent. They asked me if there was any job at the firm I would be interested in doing. So I pitched the same idea BARRA had turned down—using quantitative models to do proprietary trading. Very sensibly, they asked if I had any experience trading. Aside from a little speculation in stock index futures, I hadn't traded at all. So I played up the poker playing experience. I guess they bought it, because they offered me a job creating a proprietary trading group.

I was guaranteed a nice multiple of my BARRA compensation for each of the next two years. This was pretty good, because it was already May and I didn't have to start until September. Wall Street paid a lot better than consulting! I took the summer off, went to Hanalei, Kauai, one of my favorite places in the world, and hung out for six weeks, getting ready for the next phase of my career. I also tried to persuade my girlfriend to move to New York. She declined, and although I was disappointed at the time, it was the right move for both of us.

I had met one person at Morgan Stanley, Derek Bandeen, whom I thought had a reasonable grasp of what I hoped to accomplish. Everyone else was probably thinking, smart guy, maybe he can figure something out, and in any case he's a cheap option. I was thinking, I hope I can figure out this trading stuff, but if I don't, I'll make a huge amount of money and go back to California richer and wiser. It was slow to dawn on me that what I considered a huge amount of money, Morgan thought of as rounding error.

Derek ran prop trading in Equity Derivatives, and he was going to be my boss. Before I left for Hawaii, Derek asked me if there was anything I needed. I sent him a one-page letter describing all the data (Compustat, FACTSET, NYSE TAQ, IBES, etc.) I would need. I also asked for two offices, one for me, and one for a few programmers I wanted to hire. He told me he would try to have everything set up for me by the time I got back.

I went straight from Hanalei Bay to Times Square, about as brutal a cultural shock as I can imagine. When I showed up at Morgan Stanley and asked for Derek, I was taken to the trading floor. He was on the phone, but I spotted the letter I sent him at the bottom of a large stack of papers. It didn't take long to figure out that getting my data ready had not been a high priority—why would it be? I was an unproven commodity, and there were lots of things people could be doing to make money. Derek apologized and also told me there was no office ready, but that his boss, the head of Derivatives, was kind enough to let me use his desk on the trading floor. It was made clear that this was a pretty big deal.

I sat down and was instantly overwhelmed. People were yelling across the floor, there were monitors everywhere, many hundreds of people were packed together with a little bit of elbow room. I started to panic. How did anyone think in this environment? I called a friend of mine, Tom Cooper, who worked at GMO in Boston and had previously been fairly successful at Goldman Sachs, and asked him how he survived in a brokerage floor environment. While talking on the phone, the woman next to me, who was already talking on one phone, said, "Excuse me, I need that phone," and grabbed the receiver from my hand and ear. I was still in shock a minute later when she apologized and told me she had needed the phone to do an arbitrage trade between markets in Chicago and Tokyo. I'm still in shock thinking about it 15 years later, but it was a quick and useful introduction to the culture.

Things got better from there. I got an office built out for me, along with a bay to put a few programmers. It was a bigger request than I realized. Researchers who needed offices were kept on a separate floor where it was much quieter. A trader who needed quiet was an anomaly, but Morgan was accommodating, and I got my office.

My office was on the 33rd floor of 1251 Broadway, Morgan Stanley's headquarters at the time. Next to me were three department heads, each of whom ran very large departments of salesmen and/or traders. The day I moved in, this midlevel salesman came in and started looking at my bookshelves, walking around the place like he owned it. He turned to me and said, "Who the F are you and why the F do you get an office?" I don't remember exactly what I said, but I made sure my reply contained a lot of F-words, too. It worked. He accepted me and left.

The most important thing about Morgan Stanley was that it was a meritocracy. If you made money you got respect and power. If you didn't, you needed to make yourself useful to someone who made money. I chose to try to make it myself.

It took a couple of years for me to figure out what to trade and how to trade it. My path to success was not straightforward and involved a lot of missteps. Two of the first three people I hired didn't work out. Derek was transferred to London two months after I started. My new boss, who ran all Derivatives, left in major shakeup just as we started making money.

But I persevered. I learned from my initial mistakes and hired a number of people, many of whom I am greatly indebted to for our success. My group, Process Driven Trading, became one of the most profitable trading groups on Wall Street. We had tremendous fun building it, and in the process, created a lot of wealth for the firm and for ourselves. That's a story I hope to share in the future, but it's more "How I Became a Trader" than "How I Became a Quant."

Chapter 23

Andrew J. Sterge

President, AJ Sterge
(a division of Magnetar Capital, LLC)

When did I become a quant? Who knows? I have never once used this word to describe myself. The only time I paid any significant attention to it was after being asked to write this chapter. One thing is for sure, contrary to the title's connotation, becoming a quant is not a destination but a process. If you think you have come to the end, then you are not a quant anymore.

Another thing for sure, I never planned to do this at any early age. I went to college to play tennis, which is what I figured I would do for a living. It was not until after my third year in graduate school that I figured out it was all not going to work. I am markedly better now at getting out of losing trades. In all seriousness, however, my college and graduate school years were very formative and deserve due attention.

I chose Wake Forest University as my tennis camp—I mean college. One thing about Wake Forest: Unlike rivals Duke, University of North Carolina, or North Carolina State, Wake Forest is no party school. If you are there you might as well study, and study I did, majoring in both mathematics and English.

Wake Forest is a wonderful place. Though small and gorgeous, with a liberal arts focus, it had (and has) an incredible math department and curriculum, from which I benefited to no end. From my first semester sophomore year, I took graduate courses. In short order I completed those, and through the selfless devotion of others, most notably Professor Marcellus Waddill, I was able to continue one-on-one directed studies beyond Wake's graduate program. When I finished and got to graduate school I was in the almost exclusive company of Ivy League scholars, all of whom were brilliant. I was clearly the least such. But I knew the most math, and I completed my PhD the quickest, which unnerved everyone. I have Wake Forest to thank for it all.

It is a little funny looking back. I sit here now knowing that it all worked out. However, it was not always this way. I have the impression that if you are good in something like math, you know it very early. I, however, was no child prodigy. The only inkling that I had any math talent was in ninth grade geometry, where I was forever embarrassing my teacher with conjectures and proofs of things not in the book. The ironic thing about math is that geometry is what math is really like when you get more advanced. However, it is a single course taken in ninth or tenth grade. Math talent, however, is evaluated on one's algebra and calculus performance. I made too many mistakes to be very good at these subjects.

Bit by bit in college, though, I began to figure out that this math thing might work for me. It all had to do with problems I could solve that others could not. I remember one very specifically: Let G be a group whose order is not divisible by 3. Suppose that $(ab)^3 = a^3b^3$ for all $a,b \in G$. Prove that G must be abelian.[1] I was no childhood prodigy, but solving this problem gave me the confidence I needed that I might have a competitive advantage in this subject after all.

Graduate school was an altogether different matter. I went to Cornell—still thinking I could play tennis for a living—and was surrounded by brilliant students. What surprised me, however, was how advanced my training was. As an undergraduate I had already studied

Galois Theory, done all the problem sets in Walter Rudin's green book, *Real and Complex Analysis*,[2] and had two years of point, set, and algebraic topology. Everyone was smarter than I, but I had nothing new to learn from Cornell's core graduate math curriculum.

Then I benefited from a fortuitous incident. I came upon an evaluation of my entering graduate school class carelessly left on the copy machine. In it were intimations of doubt that I could successfully complete the PhD program. I had an immediate "I'll show them" attitude. Thank you, whoever did that.

As I said, the math core offerings were nothing new, so I looked around. Then came another fortuitous break: my discoveries of game theory and of my eventual thesis advisor, William F. Lucas. This guy is a math legend, though unlike most, as humble a man as you will ever meet. What made him a legend was an elegant counter example to John von Neumann's and Oskar Morganstern's conjecture that all cooperative N-person games have solutions according to their self-proclaimed definition of such. Until that time, cooperative game theory was thought uninteresting with no open issues. However, once Lucas proved the case was not closed, the whole subject blossomed with theories of solution concepts, some of which have proved extremely valuable in applications such as voting analysis and fair division.

Cooperative game theory is a mix of geometry and finance that gets very little attention in the mainstream math crowd. But for me it was exactly what I wanted, and I pursued it with a passion. I completed my PhD only two years from the start of Lucas's introductory class, and wrote a thesis on what relative prices you would pay for the commitments of individuals in a voting body. This is otherwise known as bribery.

My pursuit of game theory led me to another individual who has had a remarkable influence on my career, as well as to a 20-year strong friendship. That individual is Robert A. Jarrow. When I was looking around for things to do at Cornell that had anything to do with math, I was naturally led to finance, and this led me to Jarrow and his course: "Options, Futures and Forwards," or something like that. I literally could not tell you what a dividend was, but, like game theory, the subject of options pricing turned me on from the first 15 minutes in that class, and I never looked back. Both game theory and options are perfect examples of math applications in action. Math departments struggling to obtain students to major in the field should think about offering these motivational subjects.

One last topic related to Cornell, since it will come up again. Like many graduate students, I was as poor as could be. New York had a five-cent deposit on soda cans, and I remember searching through trash buckets for cans so I could collect enough to treat myself to McDonald's. A big part of becoming a quant was the hunger I had to improve my life, both literally and figuratively. Every time I go to McDonald's now I think of collecting those messy, sticky cans. I am past that now.

On to the Real World

My finance career proper started at CoreStates Financial Corp in Philadelphia. There I was, a newly minted math PhD—one who had never used a computer before, mind you—sitting in front of screens galore as the bank's head options trader. I remember not knowing the difference between a bid and an offer, and getting confused every time it was explained to me. Why was there not just one price to buy or sell, I wondered?

My three years at CoreStates were unremarkable but formative, as it was there my inquisitiveness took over and I formulated the many curiosities that have structured much of my career. It was a conservative, dull place, as very profitable regional banks can be, but there were some very good people there. One from whom I learned a lot (including the difference between bids and offers) was my boss, Gary Cantwell. Gary was an old-school bond trader who kept hand-drawn (torn but taped together) point and figure charts of the on-the-run bonds going back to the 1960s taped up on the walls of our trading room. To him, trading was all about developing gain or loss expectations and having the discipline to stick with them. Profit expectations going into a trade had to be identified beforehand, as did where you would get out. Gary may have been born a Catholic, but his real religion was "Cut Your Losses." I was a naïve but eager disciple. His lessons I think of everyday, and have taught them myself to at least a hundred others.

Gary was an avid reader and market historian and he got us all reading these neat old books published by early-twentieth-century market operators. *Reminiscences of a Stock Operator* by Edwin Lefèvre was a given. But we also got into more unusual works like Philip Carret's *The Art of Speculation, Tape Reading and Market Tactics* by Humphrey B. Neill, and several by Richard D. Wyckoff, such as *Wall Street Ventures* and *Adventures Through Forty Years.*[3] Many years later, running Cooper Neff, I imitated Gary's approach with my employees. My favorites are Ayn Rand's *The Fountainhead* and *Atlas Shrugged,* and *The Disciplined Trader: Developing Winning Attitudes,* by Mark Douglas. The Rand novels, in particular, are awe inspiring, and portray what it really means to take responsibility for one's actions—paramount for any aspiring trader or quant.

Cooper Neff

My start at Cooper Neff was inauspicious. It was raining the day of my interview, and it was a six-block walk from CoreStates at Broad and Chestnut to Cooper Neff's headquarters at the Philadelphia Stock Exchange. I arrived in my suit, soaking wet. Rich Cooper wondered why I had not taken a cab. I thought best not to tell him it would have cost 100 cans.

Options trading firms like Cooper Neff, O'Connor, and CRT were known for their grueling, tricky interviews. Mine was no exception, but I nailed it with insight into a problem I had thought deeply about, and I was lucky enough to get asked about it. It was not so much a trick probability question but more of a puzzle that Cooper, Neff, and the rest of the finance world had struggled with. It went something like this:

> Consider the USD/DM exchange rate, trading at 1.00. Let's say it can go to 1.50 or 0.50, with equal probability.[4] So fair value is 1.00 dollar. Now look at it from the DM/USD perspective: it can go to 0.67 or 2.00, for an expected value of 1.33 DM. So, its fair value is 0.75 dollars. What gives?

Though I did not know it at the time, this problem has vexed finance professionals and academics for a long time and even has a name: Siegel's Paradox.[5] Some very esteemed practitioners and academics have written about obtaining the ostensible free curvature you get as a dollar investor investing in foreign currency. Call me nuts, but I told Cooper and Neff that I saw no paradox, or at least no positive expected value trading opportunity. The USD/DM and DM/USD are priced in different units of measure. If there is no purchasing power arbitrage, then there is no edge or paradox here. I went on to explain that the problem was thinking there were two instruments, trading in a single currency, trading at prices X and $1/X$. If there were, there would indeed be free gamma everywhere. But foreign exchange rate pairs do not fit this paradigm.

Remember, this was the late 1980s; no one had yet coined the term *quanto*.[6] Nevertheless, the reason I had paid attention to this so-called paradox was that I noted how butchered Jensen's Inequality had become in the fixed income literature. In particular, the careless interchanging of the expectations operator on interest rates and bond prices alike. This is not the same error as is highlighted in the exchange rate example, but it is similar enough that I could make the connection.

Evidently Cooper and Neff had asked dozens of others this question, and no one had a clue. My confident explanation was all I needed to become director of options research in what was certainly one of the coolest and most talented organizations in the financial markets.

As an aside, only a few years after my interview things in the X, $1/X$ coexistence got interesting. The CME, for example, had the bright idea of introducing currency cross-rate contracts, like DM/JPY, but pricing them in dollars. We were all over this. However, much to our disappointment, these contracts did not succeed. Talk about free gamma!

Early Days at Cooper Neff

My interactions over many years with Richard Cooper and Roy Neff are basically what made me a quant. They are extraordinary individuals; bright as can be, funnier than hell, and chock full of insight into where to find, and how to exploit, an edge in a trading context. They are also friends from childhood but as different personality-wise as could be, a prototypical Mr. Outside and Mr. Inside, if ever there were such. Roy is the logical one, and what I learned from him about making dollars out of theory has proved immeasurable.

I guess most quants are derivatives specialists. This is not true in my case, however, as I got immersed in the idiosyncrasies of market microstructure frictions and how they could be exploited. Still, in my early days at Cooper Neff we worked on some fascinating options modeling. At least it was fascinating in the early 1990s.

Option modeling in those days was Black-Scholes based, though the market knew a lot about fat-tails, and so *winger* options would trade above Black-Scholes values. Everyone also knew that option volatility, an important parameter in the Black-Scholes model, moved around like crazy. In those days, the market concentrated on measuring, interpreting and forecasting at-the-money option implied volatilities. Although matter-of-fact today, we recognized something was amiss here. Volatility moving around was a given. At-the-money options, however, have zero volatility curvature. But out-of-the-money options do not. [Out-of-the-money options, hedged with at-the-money options, were therefore sources of free gamma, unless the out-of-the-money options were priced to reflect this gamma.] Indeed, there are two sources of gamma in an option: price gamma and volatility gamma. Naturally, there also are two sources of decay. Prices of options reflect the value of their gamma. Price gamma gets reflected in Black-Scholes values; volatility gamma does not. Unless out-of-the-money options contain an extra premium for volatility gamma, they are mispriced. This was a huge source of profits for those who figured it out. Naturally, you paid most attention to those out-of-the-money options where volatility curvature was greatest.

Another thing that really interested me early on at Cooper Neff was how options decayed. I felt it was not a smooth function of $\sigma\sqrt{t}$ as everyone seemed to think. My thought was that options decayed due to the resolution of uncertainty about where the underlying price would be at maturity. The resolution manifests itself in a narrowing of the expected distribution of returns. This narrowing may be monotonic, but it could hardly be expected to be homogeneous.

The issue became clear to me in one conversation with Roy Neff where he wondered why the term structure of volatility was so often abnormally high for the shortest-to-maturity expiration. Thinking about decay as the resolution of uncertainty provided a solution. Resolution of uncertainty is a function of information arrival. This is relatively high following a macroeconomic announcement, or, in the case of stock options, following a company-specific announcement. Information arrival is also higher at the open and close versus the middle of the trading day. And, of course, there is much more information during the trading day than overnight or over weekends. All these things make options decay more rapidly, or more slowly, than assumed by the smooth decay function embedded in the Black-Scholes model. Being blind to the heterogeneous process of option decay makes you believe option implied volatilities are more volatile than they really are.

Active Portfolio Strategies

Cooper Neff had two incarnations: first as an options market maker on exchanges around the world, next as a technology-driven, quantitative modeling firm trading equities at unheard-of high frequencies. The inflection point was in 1995, the year Cooper Neff was acquired by French bank BNP.

Active Portfolio Strategies, or APS, was our version of equity statistical arbitrage, but no one ever used those terms at the firm. To us, equity *stat arb* meant pairs trading or exploiting the residuals of an equity factor model, and nothing we were doing had anything in common with these strategies. So we made up our own name.

The genesis of APS came, oddly enough, while I was working at CoreStates in 1987. I came upon a barnburner of a book called *The Microstructure of Securities Markets*.[7] This out-of-print work is a thin but dense and abstruse book that to me was a veritable gold mine. Not so much for any specific nugget of mathematics, but more for the amazement value that someone had thought about things like supply/demand functions at the bid/offer spread level and modeling the probability of a limit-order execution. Everything I had seen in the finance literature up to this point searched for market anomalies using closing prices or assumed continuous, single-price processes. And yet, this was not the way the world worked, I thought. Stocks trade in a double-auction framework. A trade results from someone hitting a bid, taking an offer, or two sides agreeing in the middle. So looking at how stocks moved short-term meant studying market frictions and the discreteness of how stocks moved from bid/offer to bid/offer. As far as I could tell, no one had studied this before. There was no box; so thinking out of it was all one could do. It was then I figured out how I wanted to make my mark.

The story of APS seems to me a neat lesson in how quantitative finance can evolve from humble roots. We started in 1991. Five years into the process we were running regressions with tens of millions of data points. But back then there was no tick data whatsoever. The idea of creating a model of short-term expected movements of bids and offers would have to rely on nothing more than what you could currently see on a screen, without the benefit of anything like a backtest. Therefore, you had to develop a model based on your beliefs about the value of what was contemporaneous, and available. Eventually, we collected our own tick data, which in those early days was an incredible asset and competitive advantage. Nevertheless, I think it proved invaluable starting from scratch with limited resources, as it forced me to think intuitively rather than fall victim to the siren song of statistical data mining.

Here is what I mean by developing a microstructure model without data. In 1991, U.S. stocks traded in one-eight-wide minimum increments. Let's say a stock was quoted at 20 to 20 1/8. That's a 60 basis-point-wide bid offer. The last trade being on the bid versus being on the offer said something very different about that stock's likelihood of going up or down in the future. How could one quantify this without relying on historical data? It was a lot of fun. No data, no computer even—just pen and paper. I remember making ample use of hyperbolic tangent functions to represent the value of information without having to worry about my estimates of expected returns getting too far above the bid or below the offer.

Modeling stocks in a microstructure framework, for the purpose of earning a profit, is all about modeling market frictions. Bid/offer spreads are a market friction; so is the fact that markets have to open and close and that you cannot trade in unlimited size. This was hardly sexy stuff in the early 1990s, when all the rage was customized curvature in the form of structured derivative products or over-the-counter options. At that time, quantitative finance was all about improving upon Black-Scholes. I loved being apart from the crowd.

You cannot make yourself a successful quant without implementation. Although it might seem matter of fact, the greatest research in the world does no good if it cannot be implemented. And effective implementation requires a lot of courage on the quant's part. What do I mean by this? A quant must take responsibility for his work and live the P&L

day to day. Handing a model off to a trading desk to then go work on another project is a subtle way a quant tries to gain a free option: Take all the credit if the model works, blame it on traders if it does not. Working every day with traders, making their pain and the firm's pain your own, and putting yourself at risk for possible failure, is your only hope if you want to be really great in this business.

APS is an extraordinarily successful strategy. It may not be how I became a quant in the first place, but it certainly is why anyone cares. However, if I did anything right, it was not so much the math but the ability to find great people and get them to work together for a common purpose. In the beginning all you have is a formula for expected returns—an unusual one, perhaps, but still just a formula. As I am fond of saying, you cannot just take your backtests to a bank and ask them to give you the money. Implementation takes a Herculean effort from many people.

For example, perhaps the thing we did best at Cooper Neff was thinking about the trader/machine interaction. We knew high-frequency equity trading was a high-stakes videogame, and we needed the best ergonomics and the least latency possible. Roy Neff was the architect of the original APS trading system; it is a tribute to his insight and skill that it remains essentially unchanged 15 years later. It is not something that could have been designed by system engineers alone; it was Roy's skill as both trader and quant that was most salient. His design of our system was a huge key to our success.

No discussion of APS, and my role within it, would be complete without mentioning BNP, and Cooper Neff's primary liaison there, Yann Gerardin. Big-bank deals get much maligned by traders and quants, oftentimes for good reason, but ours was a home run. Under the Cooper Neff partnership, APS was a very successful strategy that made money almost every day, but we could only support a relatively small book. To fully exploit the good model we had, we needed the infrastructure that only a big bank could provide. But you need even more than that. You need a beneficent face to that organization—someone to look out for you and marshal the resources you need. Yann Gerardin was this person for us. Working tirelessly for our benefit, he sorted through the maze and got us the capital, exchange access, financial support, and other tools to grow APS 100-fold from its roots. You cannot be an effective quant in a big organization without someone like Yann looking out for your best interests.

How I Became a Quant

For me, becoming a quant has been a team effort. I have succeeded from the generosity, insight, and encouragement from a huge number of incredible people. Without their influence I would be nowhere.

In addition, and upon reflection, there are a few other things that got me to where I am. For one, I had an aptitude for difficult, abstract mathematics, and a desire to apply that talent to the financial markets. I had observed that, back when I was starting, no one was applying anything other than rudimentary math to trading. I sought to change this. I believe that while there is a lot of efficiency in markets, there is no god determining fair prices. Prices are determined by agents acting in their singular self-interest. The only way anyone knows a fair, equilibrium price has been established is for prices to overreact in both directions. I have set out to profit by making this process more efficient. In the two areas I really know anything about, options pricing and market microstructure, I read virtually every book and article published on those subjects and then find things to exploit that are not written about. Lastly, no model or mathematical insight can make money on its own. One needs a team that collectively understands every aspect of running a trading business. Becoming a quant is not an individual sport.

Chapter 24

John F. (Jack) Marshall

Senior Principal of Marshall, Tucker & Associates, LLC, and Vice Chairman of the International Securities Exchange

Within the financial engineering community, I am perhaps best known to many as the person who gave definition to the field. My effort to define financial engineering began when I undertook the challenge of writing a book, with Vipul Bansal, titled *Financial Engineering.* (The publisher, a division of Simon & Schuster, later added the rather arrogant subtitle *A Complete Guide to Financial Innovation.*) We began this book in 1988 in response to our realization that finance, as we knew it and as we had been teaching it, was undergoing profound and fundamental change.

The first step in the writing process was to interview people we considered financial engineers. I recall so many interviews that began with the interviewee asking, "Financial what?" That book effort gradually led us to recognize that there was a need for a professional society to "foster the emerging profession of financial engineering." This, in turn, led me, in 1991, along with Vipul and Robert J. Schwartz, to organize the International Association of Financial Engineers, Inc. (IAFE).

We officially launched the organization on January 1, 1992, with 40 founding members, many of whom were the people we interviewed in the writing process. I served as the executive director, Vipul served as the treasurer, and Bob served as the chairman of the board. During my tenure as the IAFE's executive director (1991 to September of 1997), the organization grew in prestige, and financial engineering, as a profession and as a discipline, grew in stature.

I have actually started in the middle of my story, so let me turn back the clock to the beginning.

From Premed to Derivatives

I was born in Brooklyn, New York, but moved to Queens at the age of seven. My father had built a small textile mill into a successful business despite intense competition from overseas. During my adolescence, he subscribed to three newspapers: The *New York Times*, *Wall Street Journal*, and the local paper, the now defunct *Long Island Press*. As a kid, I was only interested in the latter because it was the only one that contained what we called the funny pages.

When I turned 12, my father decided that it was time for my "business education" to begin. His method was simple. Every day he would circle two stories in the *Wall Street Journal*. Only after I had read them, and he had listened to my synopsis of them, could I read the funny pages in the *Press*. So every day from the age of 12, I read a little business news. Nevertheless, business did not really interest me.

I attended Archbishop Molloy High School from 1966 to 1970. I loved science and I enjoyed math, so I suppose it was not surprising that I majored in science when I got to college. I attended Fordham University, where I earned my bachelor's degree in three years. I had intended to go to medical school and, thus, was a premed student. I majored in biology and minored in chemistry.

It is often said that life takes odd turns. Mine has taken many. The first occurred after my second year at Fordham. My grandmother, who lived in Chicago, became quite ill, and I went out there to spend my summer with her. I visited her each day in the hospital and spent the rest of the day alone in her apartment. I did not know a soul in Chicago and quickly became bored and lonely. Upon learning of this, my father contacted an old college friend of his, who was a soybean trader at the Chicago Board of Trade (CBOT). He invited me to the floor to see what futures trading was all about.

I stood by his side for one full trading session and, by the end of the session, I was determined that I wanted to trade futures, too. He was a local.— Throughout the day, he bought and sold hundreds of thousands of bushels of soybeans. At the end of the day, I asked him what he intended to do with all the soybeans. He replied, "I have been trading soybeans for 31 years and I have never seen one of them suckers yet." At that moment, I understood that the soybeans weren't real (at least if you did not take delivery). I was fascinated with futures trading from that day forward.

Upon returning to New York, I informed my father that I was not going to go to medical school when I graduated. When he asked what my new plan was, I said I wanted to trade imaginary stuff. I explained that a futures contract represents an exchange of promises. One party promises to accept future delivery and the other party promises to make future delivery, but neither party intends to keep their promise as both intend to liquidate their positions via offsetting transactions. In other words, a futures contract is an exchange of promises that neither party intends to keep and both parties know it. Not surprisingly, this left my father rather perplexed.

Frustration with Academia and the Birth of a Profession

Upon returning to Fordham in the fall, I completed my degree in science, but never pursued science after graduation. Instead, I began trading commodities. From 1973 until 1980 I traded silver and gold futures. However, acutely aware of my lack of a formal business education,

I began working on an MBA in finance in 1975. The program was strictly an evening program, which I completed in 1977. Around the same time, I was accepted to a doctoral program in economics at Stony Brook University (a research campus of the State University of New York), and later was awarded a dissertation fellowship by the Center for the Study of Futures Markets at Columbia University. I defended my dissertation on September 30, 1981. My dissertation looked at the impact of technical trading on futures price volatility.

I began teaching economics and finance at a small liberal arts college while still a doctoral student, and did so for two years. In September 1981, I began a tenure-track position as an assistant professor in the Economics and Finance Department of the College of Business at St. John's University in New York. Because I had taught for two years prior to joining the St. John's faculty, I elected to come up for tenure after five years rather than the usual seven. This meant I had less time than most to build a track record worthy of promotion and tenure. I soon learned that politics plays a big, and mostly negative, role in academic life. I never learned the rules of that game, preferring to confront obsolete thinking rather than going along to get along. This was a source of repeated disappointment and frustration for me.

During my early years at St. John's, I focused my research on what I knew best—futures markets. I participated actively in the CBOT's research seminars, and began writing my first book, *Futures and Options Contracting: Theory and Practice*. I spent all my spare time researching for the book. I didn't mingle much with other faculty members because not one of them shared my research interests. The term *derivatives* had not yet been coined. I tended to prefer the company of my former colleagues from the world of commodities, and later, as I expanded into over-the-counter derivatives, my colleagues in the banking community.

I was always perplexed at the limited attention paid to derivatives by business schools throughout the 1980s and was vocal in this respect at department meetings. But my complaints always fell on deaf ears. I wanted to introduce a course in futures and options, but the curriculum committee refused to see the value of such a course and turned it down.

By the time I came up for tenure, I was clearly not regarded by the department's personnel committee as among the favored few. Indeed, they reached a split vote on my tenure in 1986. I was, however, saved by the business school dean, John C. Alexion, who saw the growing importance of the derivatives markets and who believed my research had value. John shepherded my tenure through the college personnel committee and, eventually, through the university personnel committee, but not without a major battle with my department chair. I recall one member of my department who served on the university personnel committee telling me that my tenure application did not have a chance because all my work was in derivatives and everybody on the faculty but me knew that derivatives were "only a passing fad."

During these early years as an academic, I began to consult. My consulting engagements were always derivatives related. My first engagement was with Kidder Peabody beginning in late 1981 and involved the pricing of stock index futures. I stayed with Kidder until 1985. In late 1985, I began working, as a consultant, with the swaps desk at Chase Manhattan Bank. A few years later, this brought me into contact with Bob Schwartz. On the swaps desk, I got to see a simple but amazingly versatile product, the uses of which we were only just beginning to understand. Later, in 1987, I began a consulting engagement with the First Boston Corporation that lasted until 1991. This, too, was in over-the-counter derivatives. It was during my time at Chase and First Boston that I began to appreciate just how fundamentally finance was changing, and the role that financial innovation in general, and derivatives in particular, were playing in that rapid transformation. Indeed, what I was seeing during the day in the Street was growing increasingly at odds with what I saw being taught in business schools.

Around 1987, while still at Chase, one of my MBA students, Ken Kapner, asked if I would oversee his master's thesis. He said he wanted to write his thesis on swaps. Knowing how little had been published on the subject, I cut a deal with Ken. I would oversee the thesis on the condition that we would publish it as a book when it was done. This we did in 1989, when we published *Understanding Swap Finance,* the first book on the subject. It quickly became a hit, and we eventually expanded it into a tome called *The Swaps Handbook.*

Around this time, I read an article by John Finnerty titled "Financial Engineering in Corporate Finance."[1] This article gave, for the first time to my knowledge, a name to what I was witnessing, and I embraced it. I then approached Vipul Bansal, an assistant professor in my department and one of the few who recognized that a new way of thinking was catching on. We then set out to write the book I opened this chapter with, and that, in turn, led to the formation of the IAFE.

The IAFE and the Road to MSFE Degrees

My years on Wall Street gave me a perspective that was relatively rare among finance faculty. During these years, I encountered many of the best quants I would ever meet. They were coming into finance (mostly on derivatives desks) from once alien disciplines. They included physicists, mathematicians, electrical engineers, astrophysicists, and other geeks (at least that's how most of the old timers on Wall Street saw them).

Through the IAFE I learned that I was not alone in my belief that most of academia was missing the great transformation that was taking place in finance. Other professors, here and there at universities around the world, were as frustrated by obsolete curricula as was I. They came to the IAFE as a place they could share a common love of quantitative finance at both a theoretical and an applied level. Two of those I met early on, even before the founding of the IAFE, were Anthony Herbst and Alan Tucker. They later agreed to volunteer their time to serve as the editor and coeditor of the IAFE's journal, the *Journal of Financial Engineering*. This journal served as the flagship publication of the IAFE until it was merged with the *Journal of Derivatives*.

In the early 1990s, the IAFE undertook to develop a model curriculum for degree programs in financial engineering. I tried for two years to persuade the finance faculty at my university that this really was a new discipline and there was a need for degree programs in it. With the encouragement of a sympathetic dean, I spent the next year fleshing out a detailed curriculum and course syllabi for an MS degree program in financial engineering. Unfortunately, I could never win over a sufficient number of the faculty to get the program approved. During this time, I took a one-semester visiting professorship at the Moscow Institute of Physics and Technology, which brought me into contact with some of the best engineering faculty in the world. My role was to teach finance to the faculty, who were trying to make a post–Cold War adjustment to a new Russian reality.

In 1993, George Buglierello, then the president of Polytechnic University (fondly known by its alumni as Brooklyn Poly), learned of my efforts and invited me in for a chat. He was looking for a program to bridge the old-line engineering programs of Poly with the revolution taking place in finance. Many of Poly's engineering students were going into financial services upon graduation, mostly in support roles (computer programming and related IT areas). George wanted to offer something more. We agreed to work together on a grant proposal to the Sloan Foundation, and in 1994, Sloan awarded Polytechnic University a grant of $2 million to launch the first MSFE degree program.

George invited me to join the faculty of Polytechnic as a visiting professor of financial engineering, making me the first person in the world, I believe, to hold such a title. My university granted me a two-year leave for this purpose. The first obstacle was fleshing out a curriculum that would be faithful to the IAFE model curriculum and yet exploit the strengths of the Poly faculty. The second was getting it approved by the department and then by the full faculty. The third step, the difficulty of which I had not anticipated, was getting the degree itself (MSFE) and the associated curriculum approved by the New York State Department of Education. This was complicated by the fact that the term *engineer* and its derivatives (no pun intended) is a regulated term that cannot be used lightly. We began offering elective courses for a handful of students while we prayed that NYS would see the light and approve the program, which it finally did in 1995. During this period, Andrew Kalotay joined Poly to serve as the first director of the program (handling the administrative detail and faculty recruitment) while I fought the curriculum battles.

I returned to St. John's at the end of my two-year leave just as the first of my Poly students were graduating. Over time, the program grew to become the largest graduate major at Polytechnic and is now recognized as one of the finest MSFE programs in the world.

During this period, I continued to serve as the executive director of the IAFE and encouraged other universities to develop similar programs. And, in time, they did.

My years as the executive director of the IAFE were, without a doubt, among the most rewarding of my professional life. It brought me into contact with a great many of the leading thinkers who contributed to the development of what we now call financial engineering. Among these were Robert Merton, Myron Scholes, John Hull, Emanuel Derman, Merton Miller, Franco Modigliani, Steve Ross, Harry Markowitz, Mark Rubinstein, and many others too numerous to mention.

I was introduced to several of these people by James Bicksler, one of the many unsung heroes of the IAFE. On one occasion, soon after the founding of the IAFE, the Board of Directors of the IAFE voted to extend an invitation to Fischer Black to accept a position as a Senior Fellow of the IAFE. Jim, who knew Fischer, arranged a meeting for me at Fischer's office at Goldman Sachs. On behalf of the board, I extended the invitation to Fischer, who initially declined. Soon after, however, he began attending IAFE events, apparently to see what we were all about. In 1994, the IAFE's governance committees voted Fischer the 1994 IAFE/SunGuard Financial Engineer of the Year Award (FEOY). This was the second year of this now-annual award. Fischer was most touched by the award, and soon asked if the invitation to be a Senior Fellow was still open, which it, of course, was. Unfortunately, Fischer only lived for about another year. When the IAFE later named its Foundation in Fischer's honor, his brother commented that of "all the awards Fischer received during his lifetime, the one he was most proud of was the 1994 FEOY award because it was given to him by his peers."

These people opened my eyes to so many interesting ideas. More importantly, they taught me to think differently about the problems we encounter in life, and not just the finance ones.

In closing, I would like to mention one other great experience. In 1999, I was invited to visit with David Krell, who had been secretly working on a project with Gary Katz to develop a new, completely electronic (screen-based) options exchange. This project was funded by a group of venture capitalists organized by William Porter. I had known David from both my years in derivatives and from the IAFE. I was eventually invited to serve as one of the founding members of the board of directors of the exchange they launched in 2000, the International Securities Exchange (ISE). The ISE has revolutionized the way exchange-traded equity options are traded. I have always felt that

David's and Gary's vision exemplified the "thinking out of the box" that financial engineering represents.

In 2000, I retired from academic life to concentrate my attention on my consulting firm, but I still enjoy and am forever learning from the many financial engineers who came before me and, equally, from those who are following after me. I can only dream of the new adventures that await.

Notes

Introduction

1. Joshi, Mark. *On Becoming a Quant*. Unpublished manuscript. http://www.dpmms.cam.ac.uk/~twk/Joshi.pdf. Accessed June 8, 2006.
2. Peter Bernstein, *Against the Gods* (New York: John Wiley & Sons, 1996), p. 304.
3. Emanuel Derman, *My Life as a Quant* (Hoboken, NJ: John Wiley & Sons, 2004), p. 7.
4. G.H. Hardy, *A Mathematician's Apology* (Cambridge, UK: Cambridge University Press, 1992), pp. 119–120.
5. Nassim Taleb, *Fooled by Randomness*. 2nd ed. (updated) (New York: Random House, 2005).
6. Professional traders will tell you that they love uncertainty. What they really mean, however, is that they feel they have a comparative advantage over ordinary humans in evaluating uncertain situations, that advantage being the source of their profit.
7. For many years before Black-Scholes, economists had discussed theoretical constructs for redistributing risks among market participants.

Two economists, Arrow and Debreu, each received the Prize in Economic Science in Memory of Alfred Nobel for this work.

8. Presentation Speech by Professor Bertil Näslund of the Royal Swedish Academy of Sciences, December 10, 1997. From Les Prix Nobel, the Nobel Prizes 1997, Tore Frängsmyr, ed., Nobel Foundation, Stockholm, 1998.
9. Perry Mehring, *Fischer Black and the Revolutionary Idea of Finance* (Hoboken, NJ: John Wiley & Sons, 2005).

Chapter 1

1. Much of Steven Levy's 1984 book, *Hackers: Heroes of the Computer Revolution*, takes place in the PDP-1 lab at MIT. Hacking had no criminal connotation at the time. The book is still in print.
2. It is now complete, and is utterly awesome. See the video at http://www.deltawerken.com/The-Oosterschelde-storm-surge-barrier/324.html. This is one of the premier flood control projects in the world and particularly instructive when compared with the misplaced concrete slabs in New Orleans.
3. I was the more junior of two "coleaders" on this. The big dog was one of the grand old men of the Cold War, Bruno Augenstein (http://en.wikipedia.org/wiki/Bruno_Augenstein), who was widely credited as the architect of the ICBM, and the man, who in his DoD days, signed the first check to develop the SR-71 Blackbird. He had some fine, if spooky, tales to tell.
4. LISP was the favored computer language of the artificial intelligentsia.
5. Steven was a real nice guy who gave a lot of parties. His equally nice son, Noah, was a struggling actor, working as waiter to make ends meet. Noah used to fold the napkins at Steven's parties. He worked at snazzy Hollywood restaurants and did great napkins—swans, stars, tulips, butterflies. Noah eventually got work as Dr. Carter on *ER*, so Steven farms out the napkin folding.
6. Quotron is another example of the "don't build special-purpose computers" rule. It did, and went from being synonymous with "electronic market data terminal" to being nowhere in a remarkably

short time. The first Quotrons were so alien to Wall Street types that they rearranged the "QWERTY" keyboard to be ABCDE. Schumpeter was right about capitalism being a process of creative destruction.

7. *Large* is a relative term here. The bleeding-edge machines of the mid-1980s had 32MB of memory. Fifteen years earlier, the on-board computers used on the lunar landings had 64K. Today, you can get a 1GB memory card for about forty bucks.
8. "A Little AI Goes A Long Way on Wall Street," D. Leinweber and Y. Beinart, *Journal of Portfolio Management*, Winter 1996
9. Evan's fine account of his career is in Alan Rubenfeld's book, *The Super Traders: Secrets and Successes of Wall Street's Best and Brightest* (New York: McGraw-Hill, 1995), pp. 227–252.
10. If you believe this, please contact me regarding some lucrative real estate transactions and a not-to-be-missed opportunity to help out a fine fellow in Nigeria.
11. Andre Perold and E. Schulman, *Batterymarch Financial Management* (A), (B), Harvard Business School, 1-286-113/5 (rev. 2/88).
12. For the gruesome details, see "Stupid Data Minor Tricks," *Journal of Investing*, Spring 2007. They decided to publish it as a golden oldie.
13. "Perils and Pitfalls of Evolutionary Computation on Wall Street," D. Leinweber, *Journal of Investing*, Fall 2003.
14. For details of this and other horror stories, see D. Leinweber and A. Madhavan, "Three Hundred Years of Market Manipulation," *Journal of Investing,* (Summer 2001).
15. Many of these are described in D. Leinweber, "If You Had Everything Computationally, Where Would You Put It Financially?" *Journal of Portfolio Management* (Winter 2005).
16. This crowd includes many of the same rubes who thought a machine might someday beat the World Chess Champion! HA! Can you believe these guys!!! What? Oh . . . never mind.
17. Kurzweil's ideas on machine intelligence are here http://www.kurzweilai.net/. His site, http://www.fatkat.com/, discusses his approach to investing.

18. A literary antecedent to this is found in the early days of the HAL 9000 computer from Arthur C. Clarke's novel, and Stanley Kubrick's film *2001: A Space Odyssey*. HAL, we recall, learned to sing simple songs. The singularity machine would already know them, encoded from the connections in the memory of its biological model.

Chapter 2

1. I had a long-standing interest in writing, and had won a fellowship designed to expose scientists to the mass media. That fellowship involved spending the summer of 1983 at *Newsweek* in New York.
2. Hiring me was a sufficiently risky move that BARRA insisted on hiring me as a consultant, at an hourly rate. That would make it "easier to say good-bye," according to Richard. That said, I've always been grateful to Richard and to Andrew Rudd for taking a chance on someone with an unusual background, and nothing but unproven potential.
3. Other participants in that course included Richard Lindsey and Bjorn Flesaker.
4. Rosenberg's approach modeled the impact of an embedded option as a yield spread that increased as the option increased in value. This approach fell significantly short, especially in capturing how the embedded option could dramatically reduce the effective maturity of the bond.
5. John C. Cox, Jonathan E. Ingersoll, and Stephen A. Ross, "A Theory of the Term Structure of Interest Rates," *Econometrica* 53, (1985), pp. 385–407.
6. Ronald N. Kahn and Roland Lochoff, "Convexity and Exceptional Return," *Journal of Portfolio Management* (Winter 1990), pp. 43–47.
7. We also showed that convexity strategies did relatively better in down markets than up markets. Through interactions and correspondence, Frank Jones (CIO of Guardian Life at the time) and I worked out a

more detailed understanding of these results, combining what convexity implies for parallel shifts and linear twists of the term structure of interest rates, with the correlations between shifts and twists, and what they imply in up and down markets.

8. "LBO Event Risk," Chapter 17 of Frank J. Fabozzi, ed., *Managing Institutional Assets* (New York: Ballinger, 1990), pp. 365–375.
9. Fischer Black was a key exception here, in his 1973 *Journal of Business* paper with Jack Treynor on "How to Use Security Analysis to Improve Portfolio Selection."
10. Richard C. Grinold and Ronald N. Kahn, *Active Portfolio Management*. 2nd ed. (New York: McGraw Hill, 2000). The first edition appeared in 1995.
11. Richard C. Grinold, "The Fundamental Law of Active Management." *Journal of Portfolio Management* 15, no. 3, (1989) pp. 30–37.
12. Ronald N. Kahn, and Andrew Rudd, "Does Historical Performance Predict Future Performance?," *Financial Analysts Journal* (November/December 1995), pp. 43–52.
13. Ronald N. Kahn, "Bond Managers Need to Take More Risk," *Journal of Portfolio Management* 24, no. 3 (Spring, 1998).
14. Ronald N. Kahn, "What Practitioners Need to Know . . . About Backtesting," *Financial Analysts Journal* (July/August 1990), pp. 17–20; and Ronald N. Kahn, "Three Classic Errors in Statistics, from Baseball to Investment Research," *Financial Analysts Journal* (September/October 1997), pp. 6–8.
15. At the end of that meeting, he took out a business card and pushed it toward me along the conference table. He asked for my card. I was horrified to realize that after a week's worth of client meetings, I was out of cards. I apologized profusely. But when he heard that, he slowly withdrew his card. I never collected a Fischer Black business card.
16. Richard C. Grinold and Ronald N. Kahn, "The Efficiency Gains of Long-Short Investing," *Financial Analysts Journal* (September/October 2000).

17. Ronald N. Kahn and J. Scott Shaffer, "The Surprisingly Small Impact of Asset Growth on Expected Alpha," *Journal of Portfolio Management* (Fall 2005).

Chapter 4

1. This search routine tests that an answer lies within a maximum and a minimum. It then resets one of those bounds with the average of the two. At the time, it converged on a solution with pleasing rapidity.
2. Among this group were Dean LeBaron (later to found Batterymarch Financial Management), Jeremy Grantham, Dick Mayo, and Eyk Van Otterloo (later to found Grantham, Mayo & Van Otterloo), Joan Batchelder (eventually to become head of the Fixed Income Department at Massachusetts Financial), and Len Darling (who became vice chair at Oppenheimer Funds).
3. Kalman J. Cohen and Bruce P. Fitch, "The Average Investment Performance Index," *Management Science* 12, no. 6, Series B, Managerial (February 1966), pp. B195–B215.
4. H. Markowitz, *Portfolio Selection: Efficient Diversification of Investment* (New York: Wiley, 1959), Chapter 6.
5. Refer back to the stars listed in Note 2.
6. Fischer Black's glib response to all of this was to reverse it. He suggested one way to make money was to act like a trader with information, push prices with aggressive trading. Once the prices had moved, unwind the trades using the techniques of an informationless trader.
7. Our screening involved calculating and ranking various financial ratios and rates of change. Later screening techniques would search, characterize and quantify textual information from the web and SEC documents. See David Leinweber's description of Codexa in this book.
8. The outside academic examiners included Bob Merton, Merton Miller, Stewart Myers, Steve Ross, Myron Scholes and Jay Light. Client representatives included Jon Hagler, Charley Ellis, John English, Peter Bernstein, John Bogle, Marvin Damsma, and Greta

Marshall. Fischer Black's recommendation letters, rarely more than two or three sentences long, were to the point and a delight to read.

9. Lattice was an outgrowth of MJT. MJT was founded with Bill Lupien and Murray Finebaum from Instinet and John McCormack and myself from Batterymarch. David Leinweber later joined us from Market Mind bringing humor, academic rigor and technical expertise to the firm. MJT later split. Bill Lupien went on to found Optimark, John and I were joined by Mark Hoffman to found Lattice, and David moved to First Quadrant. See his chapter (Chapter 1) in this book.
10. The chairman was John Ledyard (also head of the Department of Humanities and Social Sciences at Caltech); its president was Charles Polk, one of John's PhD students.
11. A. Perold and E. Schulman, "The Free Lunch in Currency Hedging: Implications for Investment Policy and Performance Standards," *Financial Analysts Journal* (May–June 1988), 45–50.
12. F. Black, "Universal Hedging: Optimizing Currency Risk and Reward in International Equity Portfolios," *Financial Analysts Journal* (July–August 1989).
13. Ross M. Miller and Evan Schulman, "Money Illusion Revisited: Linking Inflation to Asset Return Correlations," *Journal of Portfolio Management* (Spring 1999).
14. Ray A. LeClair and Evan Schulman, "Revenue Recognition Certificates: A New Security," *Financial Analysts Journal* (July–August 2006).
15. F. Black, Exploring General Equilibrium (Cambridge, MA: MIT Press, 1995), p. 35.

Chapter 6

1. *Raroc* is an acronym for risk-adjusted return on capital, a phrase that covers many different models for evaluating ex ante and ex post performance recognizing the risks undertaken. VaR is an acronym for value at risk, often defined as the maximum possible loss within a given confidence interval. For more information, see Wilson (98).

Chapter 7

1. The trick was so-called differentiating under the integral sign, here is the quote from *Surely You Must be Joking:* "...when guys at MIT or Princeton had trouble doing a certain integral, it was because they couldn't do it with standard methods they had learned at school...Then I come along and try differentiating under the integral sign, and often it worked. So I got a great reputation for doing integrals, only because my box of tools was different from everyone else's, and they had tried all their tools on it before giving the problem to me."
2. Of course I have to point out that Columbus did not actually discover America.
3. Fermat's last theorem states that the equation $x^n + y^n = z^n$ cannot hold for integers x, y, z if n is an integer greater than two.
4. The TRS-80 was an early entrant into the home computer market and sported a 1.77 MHz (that's 0.00177 GHz) processor, 4K of RAM, and a cassette recorder for storing programs.
5. The VIC-20 was the first "cheap" color computer. Interesting trivia about the VIC-20 include that it had approximately 3.5K of RAM for programming, used a 6502 CPU that ran at a blazing 1 MHz, and that the VIC-20 was the first computer Linus Tovalds, the developer of Linux, used.
6. For circa 1980s arcade game devotees, my favorites included Robotron, Time Pilot, Donkey Kong, Stargate, and the classic, Space Invaders.
7. Specifically, the game was modeled after Donkey Kong. For Donkey Kong enthusiasts, it was in particular based on the elevators level of Donkey Kong. The VIC-20 did not have enough memory to program more than one level. This game can be played on an IBM PC today with a VIC-20 emulator.
8. I do not remember the exact arrangement we had, but I received an advance and also a royalty agreement on each cartridge sold.
9. In the end the company's lawyers said I made it look too much like Donkey Kong and suggested some modifications, including a bomb that you had to defuse, hence the name.

10. Fermi National Accelerator Laboratory, http://www.fnal.gov/.
11. Denby, Cambell, Bedeschi, Chriss, *et al,,* "Neural Networks for Triggering," *IEEE Translation Nuclear Science* 37, no. 2 (1990), 248–254.
12. Specifically, after my undergraduate at U of C, I went to Caltech in their pure mathematics program. When I arrived at Caltech in the fall of 1989, the professor I had gone there to work with announced he was leaving after that year. With no one there to work with I applied to return to U. of C. and I also applied to Caltech's applied math program, which unlike many schools was housed in a separate department. I came very close to staying at Caltech in applied math, but in the end did my master's degree at Caltech and returned to Chicago the following fall to do my PhD in pure mathematics.
13. The paper was "Proof of the Deligne-Langlands conjecture for Hecke algebras," *Invent. Math* 87 (1987), 153–215.
14. While I was at Toronto, John was working at Goldman Sachs Asset Management managing money working for Cliff Asness, who I would later meet through John. John and Cliff along with David Kabiller and Bob Krail from Goldman Sachs founded AQR Capital Management in 1998 and it was through John that I first became interested in asset management.
15. The book was eventually published in late 1997 by Birkhauser and topped out at just over 500 pages.
16. I never published the paper but distributed it to a number of academics. In it I proposed an idea called *almost replicability* that added probability to the idea of replicating one asset's payoffs with another. I later used this idea in a paper coauthored with Michael Ong called, ironically, "Digitals Defused," which appeared in the December 1995 issue of *Risk* magazine.
17. It was actually published twice. First in Goldman Sachs *Quantitative Strategies Research Notes,* January 1994, "The Volatility Smile and Its Implied Tree," Derman and Kani, and then later in a paper called "Riding on a Smile," *Risk*, 7, no. 2 (1994), pp. 32–39. This paper and many of Derman's other papers are available from his website, www.ederman.com.

18. This was also published twice. First in Goldman Sachs *Quantitative Strategies Research Notes*, February 1996, "Implied Trinomial Trees of the Volatility Smile," Derman, Kani, and Chriss, and then in a paper by the same name published in the Summer 1996 in the *Journal of Derivatives*.
19. I published this as "Transatlantic Trees," *Risk* 9, no. 7 (1996).
20. The book was published in late 1996 by Irwin Publications under the title *Black-Scholes and Beyond*. Irwin was later acquired by McGraw-Hill publications.
21. I ended up publishing the work in the paper *Optimal Liquidation of Portfolio Transactions* with Robert Algren. *Institutional Investor* published an article about Algorithmic Trading in its November 2004 issue titled "The Orders Battle" that noted this article "helped lay the groundwork for arrival-price algorithms being developed on Wall Street."
22. Technically speaking, this is a variational problem. In particular the set of all possible paths is infinite dimensional and the optimal path for a given level of risk aversion is the solution to the Euler-Lagrange equation.
23. The final paper was "Optimal Execution of Portfolio Transactions," *Journal of Risk* 3 (Winter 2000/2001), pp. 5–39.
24. Specifically, I joined the group after Cliff Asness, John Liew and Bob Krail left to form AQR Capital Management. Ray Iwanowski and Mark Carhart became heads of the group and Ray knew of me through John Liew. John had told Ray that I was looking to move into portfolio management and invited me to interview for a role in the group.
25. One of Jim's areas of expertise and passions has been studying the volatility surface. In his later more technical teaching he produced a set of lecture notes that became widely circulated and very popular, so much so that he turned them into a book called *The Volatility Surface: A Practitioner's Guide* (Hoboken, NJ: Wiley 2006).
26. Nassim, while going on to become something of a celebrity with the publication of his book *Fooled by Randomness*, continues to teach with Jim in the program.

27. See http://www-finmath.uchicago.edu/new/msfm/home/index.php.
28. Some of the credit for this goes to the efforts of the International Association of Financial Engineers and Barry Schachter, who helped organize an Education Committee and, in conjunction with Courant, a National Financial Mathematics Recruiting day in New York. I was the first official chair of the Education Committee (http://www.iafe.org/cms_ec.html), but most of the push within the IAFE came from Barry. Today Barry and Steve Allen cochair this committee.
29. Every continent except Antarctica. The International Association of Financial Engineers posts a list of some of the mathematical finance programs (http://www.iafe.org/resources_acad.html). The website Global Derivatives has collected a more complete list of programs, see http://www.global-derivatives.com.

Chapter 10

1. Michael D. Cohen, James G. March, Johan P. Olsen, "A Garbage Can Model of Organizational Choice," *Administrative Science Quarterly* 17, no. 1 (March 1972), pp. 1–25.

Chapter 13

1. I date myself; not of course in the romantic sense but rather temporally.
2. Richard J. Sweeney, "Beating the Foreign Exchange Market," *The Journal of Finance* 41, no. 1 (March 1986), pp. 163–182.
3. The thunder machine comment is a purely personal reflection and no comment on the quality of the people at Bankers Trust—many of whom were extraordinary individuals.
4. Bankers Trust was ultimately spanked quite vigorously for some of its tomfoolery.
5. The mythical substance hypothesized during the middle ages that allowed for the spontaneous generation of life.

6. I only wish I was kidding.
7. No idea why a billion should be referred to as a "yard," but there you have it.
8. Because one thing followed the other, the first thing caused the second. I warned you about the high-brow references.
9. Where the author is right about this one . . .
10. One of these days, Rohan Douglas (the distinguished founder of Quantifi) and I will get off our barriers and finish writing the paper.
11. A point I was to learn in spades during the late summer of 1998.
12. Very few defunct trading firms continue to send performance updates.
13. Discussed this issue with my dad and he related a personal incident involving a college acquaintance by the name of Shirley Heisenberg; this is, of course, a different form of uncertainty involving the inability to simultaneously observe the position and momentum of a particle.
14. For an excellent analysis of these events see the work of Professors Brown and Steenbeek in their paper titled "Doubling: Nick Leeson's Trading Strategy."
15. Working definition here is "guy from outta town" or "guy with shiny shoes;" you take your pick.
16. A lot of traders found themselves temporarily unemployed. The ones with more significant losses managed to get book deals. Incidentally, if you're going to lose, lose big. Big losses are generally indicative of an alpha male (excuse the pun) and tend to imply that you had significant responsibility, and are therefore deserving of it again.
17. During this period I was fortunate to be working directly for the president of Nikko International, Masao Matsuda. Masao is both an organized and competent manager and a highly intellectually engaged individual, and I am forever grateful for his graceful acceptance of my somewhat eccentric behavior.
18. See the works of Douglas Martin.
19. See Richard Michaud's highly useful book *Efficient Asset Allocation.*

20. How Clintonian of me.
21. A special thanks to Sandeep Patel, Peter Williams, Stephan D'Heedene, Alvin Beh, Darko Culjak, Steve Edison, Anil Suri, Damon Wu, Noopur Gandhi, and Stephano D'Amiano of the CIA (Merrill's hedge fund quant crew) for all your positive contributions.

Chapter 14

1. The views and opinions expressed herein are those of the author and do not necessarily reflect the views of AQR Capital Management, LLC its affiliates, or its employees. The information set forth herein has been obtained or derived from sources believed by the author to be reliable. However, the author does not make any representation or warranty, express or implied, as to the information's accuracy or completeness, nor does the author recommend that the information provided serve as the basis of any investment decision.
2. I'd like to thank Brad Asness, Kent Clark, David Kabiller, Robert Krail, and John Liew for helpful comments on this draft.
3. At AQR our IT department gets a kick out of this as I often yell for help because I've lost the ability to display "Helvetica font."
4. One of these "other things" in my dissertation was a simulation study I never published that I still think is neat and an early study of what is now known as statistical arbitrage, where I concluded that it's interesting, but doesn't cover transactions costs, and then ignored several easy improvements, thereby not participating in one of the great hedge fund strategies of the late twentieth century.
5. It didn't hurt that I'd be working with my best friend Jonathan Beinner (Jon is now a Goldman partner co-running the fixed income group). It also didn't hurt that Fischer Black was then at GSAM. Working with Fischer was certainly an exciting opportunity for a 25-year-old Ph.D. student. However, Fischer did scare the heck out of me once by prominently thanking me on a paper that began with a phrase along the lines of, "Fama and French misinterpret their own work . . ." I was still writing my dissertation for Fama and French, so you can imagine my fears. Needless to say, to my relief, they took it with smiles that said, "That's Fischer."

6. Today, as the father of two sets of twins born 18 months apart, I often comment on how similar the sleep deprivation is to writing a dissertation while working at an investment bank.
7. Chiefly I mean Sharmin Mossavar-Rahmani, a Goldman Sachs partner, then and now, who was a key early supporter of the group.
8. Looking back, starting the quant group at GSAM seemed like an obvious choice, but at the time it was really scary. I was choosing to forgo academia and fixed income portfolio management, two things that both felt much more comfortable. In a nod to value investing, doing the "uncomfortable thing" worked out for me here.
9. Another consequence of taking the Goldman job was I made a final determination not to become an academic. The worst consequence of this, in turn, was that over the next decade, though he was always nice to me, countless mutual friends would say something to me like, "Gene is pretty mad at you," referring to Gene Fama's disappointment that I didn't stay in academia. Well, of course I always basked a little in the implied roundabout compliment, but I was mainly distressed. Nobody wants one of their mentors and heroes mad at them! I would always express my concern to these mutual friends, and they'd almost always say "No, he's not really mad..." or something like that, in an attempt to calm me. That was nice, but I'm too much of an empiricist not to believe the ubiquitous first comment. Happily, I can report that I think Gene is finally not mad at me (perhaps deciding I wouldn't have been that great an academic after all!).
10. Tom Dunn, a Goldman colleague at the time, definitely deserves some of the credit for this early design work.
11. If the behavioral bias explanation is correct than we would say investors were relatively optimistic (pessimistic) in Germany (France), and value works because optimism (pessimism) generally goes too far. I am sorry if my example unrealistically assumes an optimistic German.
12. We have spent a lot of years refining how we measure value and momentum for all the places we apply it, but the early measures

were really this simple (and in truth, the simple measures still capture much of the effect).

13. I have told many a client and conference attendee that I use the word *works* like a statistician, not like a real person. If something outperforms two-thirds of the years for 25 years, we are reasonably sure (ex data mining problems) it's not a statistical fluke. But if your car "worked" like I use the word, you'd fire your mechanic.
14. Another important milestone for the group was that we ended up as the engine behind the global asset allocation for Goldman's portion of the "Strategic Partnership" set up by Britt Harris then at GTE. Britt's confidence in the group was very important to us both personally and professionally.
15. An essay with quant in the title is allowed at least one science fiction reference, three if you can rhyme something with *hobbit*.
16. Goldman Sachs did not abandon our old group and today it is super successful, run by two of my hires in our last year at Goldman, Ray Iwanowski and Mark Carhart, today both Goldman Sachs partners and, of course, from the University of Chicago's PhD program in finance.
17. During our money-raising phase pre-August of 1998, one client actually used a shorthand to describe us to his board in terms of two funds they already invested in. He called us a "wonderful combination of LTCM and Tiger" referring to us as quants with a value bias. Of course, we relished the compliment as none of us knew LTCM wouldn't survive our first few months and Tiger would voluntarily close up in the next year or so as the tech bubble got crazy. In fact, many close to us have given us credit for sticking it out as compared to Tiger. That's credit I have never accepted. The world in late 1999 and early 2000 was so crazy and offensive (yeah, I'm still mad) that if, like Julian Robertson who ran Tiger, I was a billionaire, I think I would've quit, too. But, I did not have that luxury. There's a great scene in the movie *Defending Your Life* where the protagonist Albert Brooks has to point to episodes in his life where he was courageous. He points to a time he broke his leg in the wilderness and had to drag himself for miles for help. The tribunal judging him points out

to his dismay that this was not courage it was survival. Courage is running into, not out of, a burning building no matter how difficult. To this end, I am very proud of how far we dragged ourselves in the wilderness of 1999/2000, and we really did believe we were right and the world was crazy and that we were doing a great thing for our clients and ourselves by sticking to our guns—but it was not all courage, it was mainly survival.

18. Frankly I'm still mad about the whole thing. I actually carry some yellowing quotes in my wallet by hucksters during the bubble that I like to pull out at conferences. I am a great fan of the John F. Kennedy quotation, "Forgive your enemies, but never forget their names." In fact, I'm more a fan of the second half.

19. This still gnaws at me. By significantly cutting our volatility target after a very, very bad initial period, we cursed our flagship fund's track record for the foreseeable future. We did this with open eyes as it was a matter of survival (our real mistake was starting out too aggressively) and all that should matter is whether an investment is attractive going forward. But, the world does examine track records. Of course, there are many ways to slice and dice our results. For us, you could simply look at our realized returns since starting at Goldman Sachs combined with AQR, which while very inclusive, does obscure the fact that the volatility of the process was pretty different for the first four years (Goldman Sachs period plus the first year and a half at AQR). You could also look at the returns for our hedge funds and traditional portfolios we launched right after the bubble of 1999/2000, which do not contain the big volatility change effect, but is over a shorter period. You could even look at our original fund at AQR, but attempt to adjust for our change in volatility, which indicates how we would have done at a constant level of aggressiveness. All of these paint very attractive figures. In fact, just about the only way to make our live results since 1995 look less than stellar is to start the measurement in August 1998 when we started the fund at AQR and then significantly reduce our aggressiveness right after the very tough period. That this happens to be the period many in this industry focus on is what gnaws at me. Frankly, even focusing on this worst possible snapshot of us, the results, particularly when the benefits of diversification are considered, are attractive (i.e.,

they make an investor's portfolio better). However, they are not as good as they should have been, and certainly not as good as I believe a broader examination of the past reveals. Most importantly, they are not as good as I expect going forward. I will consider the next 5 to 10 years a success or failure based on how well we change this record into one that needs no explanation, however true this explanation may be.

20. I was once on a panel where I made an analogy (in direction not importance!) between people who are antiquant and racism, pointing out that people make great generalizations about how they feel about "quant" management while quants actually differ from each other a great deal (and rarely fit these generalizations anyway). When asked, tongue-in-cheek, by audience members if quants were really a race I replied, "yes, a race united solely by an inability to dance."
21. While I'm on a tour of past things I have said that I enjoy recalling, I must mention this one. At lunch once with a former actuary who was then a potential client, he said, "I quit being an actuary as it is a dying field." I responded immediately, "Yes, but at a highly predictable rate." He cracked up, and I've been vainly (in both senses of that word) repeating this story to people for years in hopes of a repeat reaction. Sadly my mortality rate on this one is quite high.
22. This might be less true for some quantitative asset allocation portfolios, like some of the strategies we run, but the general point still holds.

Chapter 16

1. The first person to settle this conjecture will win a $1 million; the $P \neq NP$ conjecture is one of the seven Clay Mathematics Institute Millennium problems. So it is (barely) possible to make money in mathematics without moving to finance! http://www.claymath.org/millennium/.
2. Conventional finance implausibly assumes that each person has a utility function and seeks to maximize it. Normative portfolio theory is concerned with the *properties* of utility functions and prescribes utility functions that achieve specified long-term goals such as maximizing the expected rate of compound return. For an excellent discussion,

see Edward O. Thorp: "Portfolio Choice and the Kelly Criterion," reprinted in W. T. Ziemba & R. G.Vickson (eds.): *Stochastic Optimization Models in Finance*, Academic Press, 1975.

3. Myron J. Gordon: *Finance, Investment and Macroeconomics: The Neoclassical and a Post Keynesian Solution* (London: Edward Elgar, 1994).
4. "You see, Daddy didn't bake the cake, and Daddy isn't the one who gets to eat it. But he gets to slice the cake and hand it out. And when he does, little golden crumbs fall off the cake. And Daddy gets to eat those." This extract from Tom Wolfe's *Bonfire of the Vanities* captures the essence of a lot of investment finance. We slice up raw financial returns and create return distributions that suit people's portfolios. Thus, for example, a collection of bonds is analogous to a side of beef, the senior tranches of a CDO are like steak, and the junior tranches are like sausages, all priced accordingly with something left over for us. Like sushi masters, we have elevated "slicing" to an art.
5. David X. Li, "On Default Correlation: A Copula Function Approach," http://www.riskmetrics.com/copulaovv.html.
6. Some of this "higher correlation in extremes" is fallacious. Boyer, Gibson, and Loretan show that the correlation of big market moves will measure higher *even if the underlying correlation that generates the returns is constant*: http://www.federalreserve.gov/pubs/ifdp/1997/597/default.htm.
7. Phillipe Jorion, "Risk Management Lessons from Long-Term Capital Management" http://papers.ssrn.com/sol3/papers.cfm?abstract_id=169449.
8. Many people *could have* written this book but they didn't. Hull's genius was to pitch it at exactly the right level for the working quant.
9. Preprint http://math.nyu.edu/research/carrp/papers/pdf/faq2.pdf, to appear in *Journal of Derivatives*.

Chapter 18

1. As I write this essay, Chester has taken a leave from his academic post at Carnegie Mellon University, in order to serve as chief economist at the Securities and Exchange Commission.

2. W. Margrabe, "An Option to Exchange One Risky Asset for Another," *Journal of Finance* (March 1978).
3. M. Kritzman, "What's Wrong with Portfolio Insurance?" *Journal of Portfolio Management* (Fall 1986).
4. See, for example, M. Kritzman and D. Rich, "The Mismeasurement of Risk, "*Financial Analysts Journal*, (May–June 2001).
5. See, for example, J-H. Cremers, M. Kritzman, and S. Page, "Optimal Hedge Fund Allocations: Do Higher Moments Matter?" *Journal of Portfolio Management* (Spring 2005) and T. Adler and M. Kritzman, "Mean-Variance Analysis versus Full-Scale Optimization: Out of Sample," *Journal of Asset Management* (November 2006).
6. William Sharpe, "An Algorithm for Portfolio Improvement," *Advances in Mathematical Programming and Financial Planning*, 1 (1987), 155–169.
7. See, for example, M. Kritzman, S. Mygren, and S. Page, "Optimal Execution for Portfolio Transitions," *Journal of Portfolio Management* (Spring 2007).
8. P. A. Samuelson,, "Risk and Uncertainty: A Fallacy of Large Numbers," *Scientia* (April/May 1963).
9. P. A. Samuelson, "Why We Should Not Make Mean Log of Wealth Big though Years to Act Are Long," *Journal of Banking and Finance* 3 December 1979.
10. See, for example, M. Kritzman, "What Practitioners Need to Know about Time Diversification," *Financial Analysts Journal* (January/ February 1994).
11. See, for example, M. Kritzman, "Are Optimizers Error Maximizers: Hype or Reality?" *Economics and Portfolio Strategy* (April 15, 2006).
12. See, for example, G.P. Brinson, L.R. Hood, and G.L. Beebower, "Determinants of Portfolio Performance," *Financial Analysts Journal* (July/August 1986).
13. R. G. Ibbotson and P. D. Kaplan, "Does Asset Allocation Policy Explain 40, 90, or 100 Percent of Performance," *Financial Analysts Journal* (January/February 2000).

14. See, for example, M. Kritzman and S. Page, "The Hierarchy of Investment Choice," *Journal of Portfolio Management* (Summer 2003).
15. See, for example, M. Kritzman, "Determinants of Portfolio Performance—20 years Later: A Comment," *Financial Analysts Journal* (January/February 2006).

Chapter 23

1. From I. N. Herstein's, *Topics in Algebra*, 2nd ed. (New York: John Wiley & Sons, 1975), problem 24, page 48.
2. *Real and Complex Analysis*, 2nd ed. (New York: Walter Rudin, Mc-Graw Hill, 1974).
3. An extensive collection of these books is still available through Fraser Publishing, Burlington, VT.
4. Don't overanalyze it; returns don't have to be normally or symmetrically distributed.
5. Jeremy Siegel, "Risk, Interest Rates, and the Forward Exchange Rate," *Quarterly Journal of Economics*, 86 (1972), 303–339.
6. Quantos are instruments that would naturally trade in one currency but are forced to trade in another. Nikkei put warrants are a perfect example.
7. Kaiman J. Cohen, Steven F. Maier, Robert A. Schwartz, David K. Whitcomb, *The Microstructure of Securities Markets* (Englewood Cliffs, NJ: Prentice Hall, 1986).

Chapter 24

1. John D. Finnerty "Financial Engineering in Corporate Finance: An Overview," *Financial Management* (Winter 1988), pp. 14–33.

Bibliography

Evan Schulman: Chairman, Upstream Technologies, LLC

Black, F. (1995). *Exploring General Equilibrium*. Cambridge, MA: MIT Press.

Black, F. (1989). "Universal Hedging: Optimizing Currency Risk and Reward in International Equity Portfolios." *Financial Analysts Journal* (July–August).

Bossaerts, Peter, Leslie Fine, and John Ledyard (2002). "Inducing Liquidity In Thin Financial Markets Through Combined-Value Trading Mechanisms." *European Economics Review* 46, no. 9 (October) 1671–1695.

Cohen, Kalman J., and Bruce P. Fitch (1966). "The Average Investment Performance Index." *Management Science* 12, no. 6, Series B, Managerial (February), pp. B195–B215.

LeClair, Ray A., and Evan Schulman (2006). " Revenue Recognition Certificates: A New Security." *Financial Analysts Journal* (July–August).

Markowitz, H. (1959). *Portfolio Selection: Efficient Diversification of Investment*. New York, John Wiley & Sons.

Mehrling, Perry (2000) "Understanding Fischer Black," http://www.econ.barnard.columbia.edu/faculty/mehrling/understanding_fischer_black.pdf.

Miller, Ross M., and Evan Schulman (1999). "Money Illusion Revisited: Linking Inflation to Asset Return Correlations." *Journal of Portfolio Management* (Spring Issue).

Perold, André F. (1988) "The Implementation Shortfall: Paper vs. Reality." *Journal of Portfolio Management* 14, no. 3 (Spring), pp. 4–9.

Perold, A., and Evan Schulman (1988). "The Free Lunch in Currency Hedging: Implications for Investment Policy and Performance Standards." *Financial Analysts Journal* (May–June), pp. 45–50.

Thomas C. Wilson: Chief Insurance Risk Officer, ING Group

Wilson, Thomas. "Is A/LM at European Life Insurers broken?" *Risk* (September 2003).

Wilson, Thomas. "Overcoming the Hurdle: Integrating Internal and External Metrics." *Risk* (July 2003).

Wilson, Thomas. "Trends in Credit Risk Management." *Journal of Financial Engineering* (December 1998).

Value at Risk, in the Handbook of Risk Management and Analysis, Volume I: Measuring and Managing Financial Risks. Ed. Carol Alexander (New York: John Wiley & Sons, 1998).

"Portfolio Credit Risk", *Economic Policy Review*, Federal Reserve Bank of New York (October 1998).

Wilson, Thomas. "CreditPortfolioView^{TM}: Technical Documentation." McKinsey & Company, 1998.

Wilson, Thomas. "Managing Credit Portfolio Risk, Parts I and II." *Risk* (September–October 1997).

Wilson, Thomas. "Credit Portfolio Risk, Parts I and II." *Journal of Lending and Credit Risk Management* (August–September 1997).

Wilson, Thomas. "Plugging the Gap." *Risk* (November 1994) (development of a delta-gamma VaR method).

Wilson, Thomas. "Debunking the Myths." *Risk* (April 1994) (application of factor analysis to VaR calculations for multicurrency term structures).

Wilson, Thomas. "Optimum Values." *Balance Sheet* (September 1993) (development of an A/LM approach for nonmaturing accounts).

Wilson, Thomas. "Infinite Wisdom." *Risk* (June 1993) (VaR implications of using estimated volatilities causing fat tails).

Wilson, Thomas, and P. Wuffli. "Asset/Liability Management." *Journal of Finanz und Portfolio Management* (Spring 1993).

Wilson, Thomas. "Raroc Remodeled." *Risk* (September 1992) (refinement of a risk-adjusted return on capital measure).

Peter Jäckel

J.D. Alanen and D.E. Knuth. Tables of finite fields. *Sankhya, the Indian Journal of Statistics, Series A* 26, no. 4, December 1964.

F. Black. The pricing of commodity contracts. *Journal of Financial Economics* 3 (1976), pp. 167–179.

P. Boyle. Options: a Monte Carlo approach. *Journal of Financial Economics* 4 (May 1977), pp. 323–338.

M. E. da Silva and T. Barbe. Quasi Monte Carlo in Finance Extending for High Dimensional Problems. *Economia Aplicada* 9, no. 4 (December 2005). www.scielo.br/pdf/ecoa/v9n4/v9n4a04.pdf.

Stephen Joe and Frances Y. Kuo. Remark on algorithm 659: Implementing sobol's quasirandom sequence generator. *ACM Transactions of Mathematical Software* 29, no. 1 (2003), pp. 49–57.

Sean Erik O'Connor. Computing Primitive Polynomials. www.seanerikoconnor.freeservers.com/Mathematics/AbstractAlgebra/PrimitivePolynomials, 1986.

I. M. Sobol'. Uniformly Distributed Sequences with an Additional Uniform Property. *USSR Computational Mathematics and Mathematical Physics* 16, no. 5 (1976), pp. 236–242.

E. Sugimoto. A short note on new indexing polynomials of finite fields. *Information and Control*, 41 (1979), pp. 243–246.

About the Contributors

Steve Allen is clinical associate professor of mathematics and deputy director of the Mathematics in Finance Masters Program at New York University's Courant Institute of Mathematical Sciences. Mr. Allen joined the NYU faculty full-time in 2004, after 35 years in the finance industry, most recently as managing director of JP Morgan Chase, in charge of risk methodology. Previous positions he has held include global head of market risk management for Chase's derivative products, director of modeling and systems for Chase's trading activities, head of modeling and analytics for Chase's Asset-Liability Management Committee, and deputy director of Management Science. He studied mathematics as an undergraduate at Columbia College and as a graduate student at New York University's Courant Institute of Mathematical Sciences. He has taught in the Courant master's in math finance program since 1998 and is the author of *Financial Risk Management: A Practitioner's Guide to Managing Market and Credit Risk,* and coauthor of *Valuing Fixed Income Investments and Derivative Securities.*

Mark Anson is the chief executive officer of the British Telecommunications Pension Scheme (BTPS) and the chief executive officer of

Hermes Pensions Management Ltd. He is also chairman of the board of the International Corporate Governance Network, the largest organization in the world dedicated to better governance of public corporations. Dr. Anson previously served as the chief investment officer at the California Public Employees' Retirement System (CalPERS). He implemented the concept of separating beta from alpha, which generated over $9 billion of excess returns for CalPERS. He received a law degree from Northwestern University and a PhD and master's in finance from the Columbia University Graduate School of Business. Mark attended St. Olaf College in Minnesota, where he double majored in economics and chemistry. Mark has also received the *Distinguished Scholar Award* from the Institute of International Education and the Fulbright Foundation, as well as the Bernstein Fabozzi/Jacobs Levy Award for his article published in *The Journal of Portfolio Management* in 2004. In addition to the *Handbook of Alternative Assets*, Mark has published three financial textbooks and more than 80 research articles on the topics of separating beta from alpha, institutional fund management, business models for the asset management industry, corporate governance, hedge funds, real estate, currency overlay, credit risk, private equity, risk management, and asset allocation. He serves on editorial and advisory boards for the *Journal of Portfolio Management, Journal of Alternative Investments, Journal of Private Equity, Journal of Investment Consulting,* and *Journal of Derivatives Accounting*, and on advisory and executive committees for the New York Stock Exchange, Euronext, MSCI-Barra International Indexes, the International Association of Financial Engineers, The CFA Institute's Task Force on Corporate Governance, The CFA Institute's Committee on Global Investment Performance Standards, The Conference Board Commission on Public Trust, The Center for Excellence in Accounting and Security Analysis at Columbia University, The Dow Jones-AIG Commodity Index Board, and the National Association of State Investment Officers.

Clifford S. Asness is the managing and founding principal of AQR Capital Management, LLC. Before cofounding AQR Capital Management, Dr. Asness was at Goldman, Sachs & Co., where he was a managing director and director of quantitative research for the Asset Management Division. He has written many articles published in the *Journal of Portfolio Management* and the *Financial Analysts Journal*. He has received Bernstein

Fabozzi/Jacobs Levy Awards for his articles published in *The Journal of Portfolio Management* published in 2001 and 2003, the Graham and Dodd Award for the year's best paper (2003), a Graham and Dodd Excellence Award (2000), the award for the best perspectives piece (2004), and the Graham & Dodd Readers' Choice Award (2005), all from the *Financial Analysts Journal*. In addition, the CFA Institute awarded Dr. Asness the James R. Vertin Award, which is periodically given to individuals who have produced a body of research notable for its relevance and enduring value to investment professionals. He is on the editorial board of the *Journal of Portfolio Management*, the editorial board of the *Financial Analysts Journal*, the governing board of the Courant Institute of Mathematical Finance at NYU, the Council on the Graduate School of Business and the university-wide Investment Committee at the University of Chicago, and the Leadership Council of the Robin Hood Foundation. Dr. Asness received a BS in economics from the Wharton School and a BS in engineering from the Moore School of Electrical Engineering at the University of Pennsylvania. He received an MBA and a PhD in finance from the University of Chicago.

Tanya Styblo Beder has built three businesses during her 20-year career in the global capital markets. Beginning in 2004, Ms. Beder designed and built as its CEO Tribeca Global Management LLC, Citigroup's multistrategy hedge. As CEO, she also was responsible for several other institutional fund offerings including convertibles, distressed debt, merger arbitrage, and credit. Beginning in 1999, she built the Strategic Quantitative Investment Division (SQID) of Caxton Associates LLC. Before her roles in the hedge fund industry, Ms. Beder founded two consulting firms specializing in risk measurement, risk oversight, capital markets and derivatives—Capital Market Risk Advisors in 1994 and SB Consulting Corp. in 1987. Before this, she was a vice president of The First Boston Corporation, where she started her career in mergers and acquisitions and was a consultant in the financial institutions practice at McKinsey & Company. Ms. Beder is on the board of directors of OpHedge Investment Services, is a Risk Management Advisor to the LongChamp Group Inc., and serves on the board of directors of the International Association of Financial Engineers. She serves on the advisory board of Columbia University's Financial Engineering Program and is a Fellow of the International Center for Finance at Yale and has served on the National Board of Mathematics and their Applications. Ms. Beder has taught as an adjunct professor at Yale University's School of Management, Columbia University's Graduate School of Business and Financial Engineering, and the New York Institute of Finance. She was an author of the *Risk Standards for Institutional Investors and Institutional Investment Managers* and has published numerous articles in the *Journal of Portfolio Management*, the *Financial Analysts Journal*, *Harvard Business Review*, *Journal of Financial Engineering*, and with Probus Publishing, John Wiley & Sons, and Simon & Schuster. Ms. Beder holds an MBA in finance from Harvard University and a BA in mathematics from Yale University.

Gregg E. Berman is responsible for Strategic Development at RiskMetrics Group. Dr. Berman joined RiskMetrics as a founding member during the time of its spinoff from J.P. Morgan in 1998 and has held a number of roles from research to heads of product management, market risk, and of business management. Before joining RiskMetrics Group, Dr. Berman comanaged a number of multiasset hedge funds within the ED&F Man umbrella. He began his hedge fund career in 1993, researching and developing multiasset trading strategies as part of Mint Investment Management Corporation. Dr. Berman is a physicist by training and holds degrees from Princeton University (PhD 1994, MS 1989), and the Massachusetts Institute of Technology (BS 1987).

Peter Carr has, for the past three years, headed Quantitative Financial Research at Bloomberg and the Master's in Mathematical Finance program at NYU's Courant Institute. Previously, Dr. Carr headed equity derivative research groups for six years at Banc of America Securities and at Morgan Stanley. His academic positions include four years as an adjunct professor at Columbia University and eight years as a finance professor at Cornell University. He received his PhD in finance from the University of California–Los Angeles (UCLA) in 1989. Dr. Carr has published extensively in both academic and industry-oriented journals. He serves as the treasurer of the Bachelier Finance Society and an as associate editor for six journals related to mathematical finance and derivatives. Dr. Carr has won the *Wilmott* "Cutting Edge Research" award (2004) and was named "Quant of the Year" by *Risk* (2003).

Neil Chriss was until recently managing director of Quantitative Strategies at SAC Capital Management, LLC. Before joining SAC, Dr. Chriss was cofounder and president of ICor Brokerage Ltd., a derivatives trading firm that was sold to Reuters Group PLC in 2004 and forms its global electronic trading system for interest rate and foreign exchange derivatives. Before ICor, he was a portfolio manager at Goldman Sachs Asset Management and worked in quantitative research in the Institutional Equities Division of Morgan Stanley. Dr. Chriss developed and served as the first director of the Program in Mathematics in Finance at NYU's Courant Institute and currently sits on its board. He has held positions in the mathematics departments at Harvard University and the Institute for Advanced Study. He is a founding member of the board of Math for America, a nonprofit dedicated to improving the quality of mathematics teaching in the United States. He is also a member of the board the Mathematical Finance program at University of Chicago. Dr. Chriss has published extensively in quantitative finance – including "Optimal Execution of Portfolio Transactions" a seminal paper on algorithmic trading, "Optimal Portfolios from Ordering Information," and the book *Black-Scholes and Beyond: Modern Option Pricing*. Dr. Chriss holds an BS and PhD in mathematics from University of Chicago and an MS in mathematics from California Institute of Technology.

Andrew Davidson is president and founder of Andrew Davidson & Co., Inc., a consulting firm specializing in the application of analytical tools to investment management. He created a set of proprietary analytical tools including prepayment and option-adjusted spread (OAS) models for fixed rate mortgages, adjustable rate mortgages, collateralized mortgage obligations (CMOs), and asset-backed securities (ABS). He has also developed the Flow Uncertainty Index, which is used by insurance regulators to assess the risk of CMO portfolios. Mr. Davidson worked for six years at Merrill Lynch as a managing director, developing analytical tools for prepayment and option-adjusted spread models and portfolio analysis, as well as trading systems for the mortgage desk covering adjustable rate mortgages (ARMs), CMOs, pass-throughs, interest-only-strips (IOs/Pos), and over-the-counter (OTC) options. Mr. Davidson is coauthor of the books *Securitization: Structuring & Investment Analysis, Mortgage-Backed Securities: Investment Analysis & Valuation Techniques and*

Collateralized Mortgage Obligations. He has written numerous articles that have appeared in the *Handbook of Mortgage-Backed Securities, Mortgage-Backed Securities: New Applications and Research* and the *Journal of Real Estate Finance and Economics.* Currently, he publishes *Quantitative Perspectives*, on issues related to the mortgage and derivatives markets, and *The Pipeline*, for participants in the fixed income industry. Mr. Davidson was previously a financial analyst in Exxon's treasurer's department. He received an MBA in finance at the University of Chicago and a BA in mathematics and physics at Harvard.

Bjorn Flesaker is a senior quant at Bloomberg L.P., where his current research is focused on the pricing and risk management of credit derivatives. Before joining Bloomberg in 2006, Dr. Flesaker spent the previous thirteen years managing quant, analytics and technology groups for a number of financial institutions, including Merrill Lynch, Bear Stearns, Gen Re Securities and MBIA Inc. Starting out his career as an assistant professor of finance at the University of Illinois at Urbana-Champaign, he has written a number of papers related to derivatives pricing, most notably "Positive Interest," which he published with Lane Hughston in *Risk* in 1996. Dr. Flesaker is a graduate of the Norwegian School of Management and he holds a PhD in finance from the University of California at Berkeley.

Bruce I. Jacobs is cofounder and principal of Jacobs Levy Equity Management, a leading provider of quantitative equity strategies. The firm, which celebrated its 20th Anniversary in 2006, serves a prestigious global roster of more than 50 clients with over $20 billion in institutional assets under management. Dr. Jacobs is cochief investment officer, portfolio manager, and codirector of research. Dr. Jacobs has written numerous articles on equity management and received the *Financial Analysts Journal* Graham and Dodd Award in 1988, a Bernstein Fabozzi/Jacobs Levy Award for his article published in *The Journal of Portfolio Management* in 1999, and an award from *The Journal of Investing* in 1998. Many concepts introduced in these articles have become industry terms, including "market complexity," "disentangling," "pure returns," "law of one alpha," "integrated long-short optimization," and "trimability." Dr. Jacobs is author of *Capital Ideas and Market Realities: Option Replication, Investor Behavior, and Stock Market Crashes* (Blackwell), coauthor with Ken Levy of *Equity Management: Quantitative Analysis for Stock Selection* (McGraw-Hill), coeditor with Ken Levy of *Market Neutral Strategies* (Wiley), and coeditor of *The Bernstein Fabozzi/Jacobs Levy Awards: Five Years of Award-Winning Articles from the Journal of Portfolio Management* (Institutional Investor). Formerly he was first vice president of the Prudential Insurance Company of America, where he served as senior managing director of a quantitative equity management affiliate of the Prudential Asset Management Company and managing director of the discretionary asset allocation unit. Prior to that, he was on the finance faculty of the University of Pennsylvania's Wharton School and consulted to the Rand Corporation. Dr. Jacobs has a BA from Columbia College, an MS in Operations Research and Computer Science from Columbia University's School of Engineering and Applied Science, an MSIA from Carnegie Mellon University's Graduate School of Industrial Administration, and an MA in Applied Economics and a PhD in Finance from the University of Pennsylvania's Wharton School. He is an associate editor of *The Journal of Trading* and on the Advisory Board of *The Journal of Portfolio Management*.

Peter Jäckel received his D. Phil. in physics from Oxford University in 1995. In 1997, Dr. Jäckel joined Nikko Securities and in 1998, he moved to NatWest, which later became part of the Royal Bank of Scotland Group. In 2000, he joined Commerzbank Securities' product development group, where he became cohead in 2003. Since September 2004, Dr. Jäckel has been with ABN AMRO as Global Head of Credit, Hybrid, Commodity, and Inflation Derivative Analytics.

Ron Kahn heads Barclays Global Investors' efforts to develop new and innovative active equity strategies. His roles at BGI have included global head of equity research and head of active equities in the U.S. Prior to joining BGI in 1998, Dr. Kahn spent more than seven years as director of research at Barra. He has published widely, coauthored with Richard Grinold the book *Active Portfolio Management: Quantitative Theory and Applications,* and won a Bernstein Fabozzi/Jacobs Levy Award for Best Article published in the *Journal of Portfolio Management* in 2006. He is on the editorial advisory boards of the *Journal of Portfolio Management* and the *Journal of Investment Consulting*. He teaches the course on Equities Markets in University of California–Berkeley's Financial Engineering

Program. Dr. Kahn received a PhD in physics from Harvard University in 1985 and an AB in physics from Princeton in 1978.

Stephen Kealhofer is the managing principal and head of research and product revelopment for Diversified Credit Investments, a manager of institutional credit funds. He founded DCI in 2003 with David Solo and Mac McQuown. From September 1988 until April 2003, Kealhofer was associated with KMV LLC. KMV was the leading provider of quantitative credit tools and data to banks, insurance and other credit investors. He founded KMV in 1988 with Oldrich Vasicek and Mac McQuown, and served as the managing partner and subsequently CEO from 1988 to June 2000. He was responsible for research and product development at KMV. KMV was sold to Moody's, the debt rating agency, in April 2002. KMV was known for its implementation of Robert Merton's credit valuation approach, which was used to provide daily estimates of default probabilities on over 30,000 companies in over 40 different countries. It was also known for developing the first practical portfolio model for credit portfolios, and implementing it at many of the world's major banks. Both the default probabilities, known as EDFs, and the portfolio models remain active product lines of Moody's KMV, the continuing business. Prior to founding KMV, Kealhofer was director of research for diversified corporate loans in San Francisco from June 1987 until September 1988. He was an assistant professor at the Graduate School of Business at Columbia University from July 1981 until June 1987, and a visiting assistant professor at the Haas Business School at the University of California–Berkeley, from July 1985 until June 1987. Kealhofer received his BA in economics from Macalester College in 1976, and a PhD in economics from Princeton University in 1983.

Mark Kritzman is president and CEO of Windham Capital Management, LLC, where he is responsible for managing research activities and investment advisory services. He teaches a graduate course in financial engineering at the Massachusetts Institute of Technology (MIT) Sloan School of Management. Mr. Kritzman serves on corporate and nonprofit boards, including those of the Institute for Quantitative Research in Finance, the International Securities Exchange, and State Street Associates. He is also a member of advisory and editorial boards, including *Emerging*

Markets Review, Financial Analysts Journal, the International Association of Financial Engineers, the *Journal of Alternative Investments,* the *Journal of Derivatives,* and the *Journal of Investment Management.* Mr. Kritzman won Graham and Dodd awards in 1993 and 2002, the Research Prize from the Institute for Quantitative Investment Research in 1997, and the Bernstein Fabozzi/Jacobs Levy Awards in 2003 and 2006. In 2004, he was elected a Batten Fellow at the Darden Graduate School of Business Administration, University of Virginia. Mr. Kritzman received an MBA in business at New York University.

David Leinweber is the founder of two pioneering financial technology firms. Clients at his consulting and software development business include some of the world's largest investment managers and hedge funds. These tasks involve trading systems and automated analysis of textual and Internet information sources. All build on his history of innovation in financial technology. At the RAND Corporation, he directed research on real-time applications of artificial intelligence that led to the founding of Integrated Analytics Corporation. IAC was acquired by the Investment Technology Group, (NYSE:ITG) and, with the addition of electronic order execution, its product became QuantEx, an electronic execution system still in use for millions of institutional equity transactions daily. Large institutions concerned with controlling transactions costs and proprietary traders found them particularly valuable. As managing director at First Quadrant, he was responsible for institutional quantitative global equity portfolios totaling $6 billion. These long and market neutral strategies utilized a wide range of computerized techniques for stock selection and efficient trading. Quantitative investing is driven by electronic information, and the Internet dramatically transformed the financial information landscape. This led to the founding of Codexa Corporation, an Internet-based information collection, aggregation and filtering service for institutional investors and traders. The company's clients included many of the world's largest brokerage and investment firms. Dr. Leinweber is a scientific advisor to Monitor 110, a web 2.0 information service for hedge funds, brokers and institutions. As a visiting faculty member at Caltech, Dr. Leinweber' worked on practical applications of ideas at the juncture of technology and finance. He is a collaborator in the Harvard Business School's e-Information project.

Dr. Leinweber has advanced the state of the art in the application of information technology in both the sell-side world of trading and the buy-side world of quantitative investment. He has published and spoken widely in both fields. He has gone five rounds against the *Wall Street Journal* dartboard, and is an editor of the *Journal of Trading*. In his misspent youth, he graduated from MIT, where he was one of the first 5,000 people on the Internet. That was when it was called the ARPAnet and wasn't cool. He also has a PhD in Applied Mathematics from Harvard. But on a good day, it's hard to tell.

Kenneth N. Levy is cofounder and principal of Jacobs Levy Equity Management, a leading provider of quantitative equity strategies. The firm, which celebrated its 20th Anniversary in 2006, serves a prestigious global roster of more than 50 clients with over $20 billion in institutional assets under management. He is cochief investment officer, portfolio manager, and codirector of research. He has written numerous articles on equity management and is a recipient of the *Financial Analysts Journal* Graham and Dodd Award in 1988 and a Bernstein Fabozzi/Jacobs Levy Award for his article published in *The Journal of Portfolio Management* in 1999. Many concepts introduced in these articles have become industry terms, including "market complexity," "disentangling," "pure returns," "law of one alpha," "integrated long-short optimization," and "trimability." Mr. Levy is coauthor with Bruce Jacobs of *Equity Management: Quantitative Analysis for Stock Selection* (McGraw-Hill) and coeditor with Bruce Jacobs of *Market Neutral Strategies* (Wiley), and coeditor of *The Bernstein Fabozzi/Jacobs Levy Awards: Five Years of Award-Winning Articles from the Journal of Portfolio Management* (Institutional Investor). Formerly he was managing director of a quantitative equity management affiliate of the Prudential Asset Management Company. Before that, he was responsible for quantitative research at Prudential Equity Management Associates. He has a BA in Economics from Cornell University and MBA and MA degrees in Applied Economics from the University of Pennsylvania's Wharton School. He has completed all requirements short of the dissertation for a PhD in Finance at Wharton. He is a CFA and has served on the CFA Candidate Curriculum Committee, the POSIT Advisory Board, and the investment board of a community foundation.

Allan M. Malz is head of Risk Management at Clinton Group, a manager of multistrategy hedge funds and certified certificates of deposit (CDO). He is responsible for risk management and capital allocation across all strategies. Before joining Clinton Group, Dr. Malz was head of research at RiskMetrics Group, which he joined on its spinoff from J.P. Morgan in September 1998. In addition to leading the research team and developing market data products, he worked on forecasting financial crises and on risk measurement for options. Dr. Malz spent his earlier career at the Federal Reserve Bank of New York as a researcher and foreign exchange trader, eventually heading the Fed open market and foreign exchange desks' efforts to improve market monitoring and analysis. His work has been published in a number of industry and academic journals. Dr. Malz holds a PhD in Economics from Columbia University, where he currently also teaches a graduate course in risk management.

John F. Marshall is senior principal of Marshall, Tucker & Associates, LLC, a financial engineering and derivatives consulting firm. Dr. Marshall is a founding director and currently vice chairman of the International Securities Exchange, and he is the CEO of Port Jefferson Capital I, LLC, a venture capital firm. Dr. Marshall is the author of several books on financial products, markets, and analytics, including *Futures and Option Contracting* (South-Western), *Investment Banking & Brokerage* (McGraw-Hill), *Understanding Swaps* (Wiley), *Financial Engineering: A Complete Guide to Financial Innovation* (Simon & Schuster), and *Dictionary of Financial Engineering* (Wiley, 2000). He has also written several dozen articles in professional journals. Dr. Marshall served on the faculty of the Tobin Graduate School of Business of St. John's University. From 1991 to 1998, Dr. Marshall served as the executive director of the International Association of Financial Engineers (IAFE). From 1997 through 1999, he served on the board of directors of the Fischer Black Memorial Foundation. From 1991 to 1995, Dr. Marshall served as the managing trustee for Health Care Equity Trust, a unit investment trust sponsored by Paine Webber. From 1994 to 1996, Dr. Marshall served as visiting professor of financial engineering at Polytechnic University. Dr. Marshall currently serves on faculty of the NASD (National Association of Securities Dealers) Institute at Wharton Certificate Program.

He earned his undergraduate degree in biology/chemistry from Fordham University in 1973, an MBA in finance from St. John's University in 1977, an MA in quantitative economics from the State University of New York in 1978, and a PhD in financial economics from the State University of New York at Stony Brook in 1982.

Peter Muller is a senior advisor at Morgan Stanley and provides ongoing counsel to Process Driven Trading, a proprietary trading group he founded in 1992. Prior to joining Morgan, Mr. Muller was with Barra for seven years. He created the BARRA Brainteaser, a monthly investment-related puzzle. Mr. Muller graduated with honors from Princeton University with a BA in mathematics. He is a member of the editorial board for the *Journal of Investment Management* and has previously served in the same capacity for the *Financial Analysts Journal* and the *Journal of Portfolio Management*. Mr. Muller has served on the advisory board for both MIT's Track in Financial Engineering (Sloan School) and NYU/Courant's Program in Financial Mathematics. He is chairman, Board of Advisors, for Chalkstream Capital, an investment fund. Mr. Muller's published research includes work on financial optimization, mortgage prepayments, and equity valuation models. Outside of finance, Muller has released two albums as a pianist/singer-songwriter, and enjoys poker, creating crossword puzzles, snowboarding, and surfing.

Leslie Rahl is the founder and president of Capital Market Risk Advisors, Inc. CMRA is the preemminent financial advisory firm specializing in risk management, hedge funds, financial foerensic and derivatives CMRA and its predecessor firms have played an integral role in the evolution of hedge funds, derivatives, structured securities, and risk management for more than 16 years. Mrs. Rahls has 35 years of financial market experience. Mrs. Rahl spent 19 years at Citibank, including nine years as cohead of Citibank's Derivatives Group in North America. She launched its caps and collars business in 1983 as an extension of the proprietary options arbitrage portfolio she ran, and was a pioneer in the development of the swaps and derivatives business. Mrs. Rahl was named one of the Top 50 Women in Finance by *Euromoney* in 1997 and was profiled in the both the fifth and tenth anniversary issues of *Risk* magazine. She was listed in "Who's Who in Derivatives" by *Risk* magazine and was profiled in *Fortune's* "On the Rise and *Institutional Investor's* "The Next Generation of Financial Leaders." Mrs. Rahl is on the board of directors of Fannie Mae and the International Association of Financial Engineers (IAFE). She is on the Investment Advisory Committee of the New York State Common Retirement Board ($115 billion) and is a member of the hedge fund committee of the Alternative Investment Management Association (AIMA). She was a director of the International Swaps Dealers Association (ISDA) for five years, chaired the IAFE's Investor Risk Committee (IRC), and recently retired from the MIT Investment Management Company Board of Directors. Mrs. Rahl is the cofounder of the High Water Women Foundation and focuses on international microfinance and intellectual capital volunteerism. She was on the board of 100 Women in Hedge Funds for its formative first three years and chaired the philanthropy committee. Mrs. Rahl is the author of *Hedge Fund Transparency: Unraveling the Complex and Controversial Debate* (Risk Books) and is the editor of *Risk Budgeting a New Approach to Investing* (Risk Books). Her articles have appeared in a wide range of publications. Mrs. Rahl received her undergraduate degree in computer science from MIT in 1971 and her MBA from the Sloan School at MIT in 1972.

Evan Schulman is chairman of Upstream Technologies LLC. Before Upstream, Mr. Schulman cofounded Lattice Trading, which was acquired by State Street Global Advisors in 1996. Lattice is an advanced alternative trading system that integrates order-matching with order-routing and connects to global computerized markets. In 1975, as director of Computer Research at Keystone Funds in Boston, he completed what is generally regarded as the first equity program trade. During the 1980s, Mr. Schulman developed computerized investment and trading systems at Batterymarch Financial Management. Currently, he is a director of Net Exchange. Mr. Schulman served in the Royal Canadian Air Force, received his BA from the University of Toronto, and his MA from the University of Chicago.

Julian Shaw is head of risk management for Permal Investment Management Services Limited. Mr. Shaw joined the Permal Group from Barclays Capital in 2004, where he was global head of market risk. Before

Barclays, Mr. Shaw was the quantitative analyst at Gordon Capital (market makers for Canadian stock index futures and options). He later joined CIBC (Canadian Imperial Bank of Commerce) as vice president for New Products on the fixed income derivatives desk. Mr. Shaw has a master's degree in mathematics and is a Chartered Financial Analyst.

Andrew Sterge is president of the AJ Sterge Division of Magnetar Capital, LLC. Prior to becoming a division of Magnetar in 2006, Mr. Sterge had founded AJ Sterge Investment Strategies specifically to manage pioneering, quantitative investment strategies that would exploit inefficiencies in the reinsurance markets. From 1989 to 2004 Mr. Sterge served at the Cooper Neff Group, including as the firm's chairman and chief executive officer from 1999. Mr. Sterge joined Cooper Neff & Associates as director of options research. In this position, Mr. Sterge developed options pricing models that captured the effects of fat-tailed and skewed distributions, as well as investors' relative risk aversion for the downside versus upside insurance aspect of options. Mr. Sterge was promoted to partner in 1993. Prior to joining Cooper Neff, Mr. Sterge was employed by CoreStates Financial Corporation where he was assistant vice president, trading interest rate options from September 1986 to November 1989. In 1991, Mr. Sterge founded a new variety of short term equity trading based on models of stock market microstructure, or how stocks' bids and offers evolve over time and in response to order flow and other information. Called Active Portfolio Strategies, this business flourished following the acquisition of Cooper Neff by Banque Nationale de Paris (now BNP Paribas) in 1995. Effectively, an internal hedge fund strategy, Active Portfolio Strategies at times managed well over $20 billion in global equity positions for BNP Paribas. Mr. Sterge's groundbreaking strategy was profiled in a December 1997 front page *Wall Street Journal* article titled "Trading by the Numbers." Mr. Sterge is a 1981 graduate of Wake Forest University, where he received his bachelor of science degree in mathematics. He was awarded the Kenneth Tyson Raynor Math scholarship in 1980 as well as the John Y. Phillips Prize in 1981, given to the senior who most excels in the study of mathematics. Mr. Sterge received his PhD degree in mathematics from Cornell University in 1985. His doctoral thesis built a mathematical model of coalition formation in a voting context. As an application, Mr. Sterge measured the relative importance of voters in the United States Federal Legislature. In addition to his academic credentials, Mr. Sterge published an article for the May/June 1989 issue of the *Financial Analysts Journal* entitled "On the Distribution of Financial Futures Price Changes."

Andrew Weisman is managing director, responsible for strategy, risk management, and portfolio analytics, at Merrill Lynch. Mr. Weisman was a founding member and director of research and risk management for Strativarius Capital Management LLC, a New York–based Global/Macro hedge fund. Until April 30, 2002, Mr. Weisman was the chief investment officer and member of the board of directors for Nikko Securities Co. International, Inc. Mr. Weisman has published an extensive collection of articles on asset allocation and risk issues related to hedge funds. Research awards include a Bernstein Fabozzi/Jacobs Levy Award for Outstanding Article published in *The Journal of Portfolio Management* in 2002, and the GAIM Research Paper of the Year in 2003. He is currently a member of the editorial advisory board of the *Journal of Portfolio Management*. Previous professional experience includes his position as head of Bankers Trust's Automated Currency Trading Unit and senior asset manager for Commodities Corporation LLC (Goldman Sachs Princeton LLC). Mr. Weisman has a BA from Columbia College (1982), a master's in international affairs specializing in international business from the Columbia University School of International and Public Affairs (1983), and has completed all course work and comprehensive exams towards a PhD in money and financial markets at the Columbia University Graduate School of Business.

Thomas C. Wilson is currently general manager and chief insurance risk officer of ING Group. Since 2006, he has also been chairman of the Chief Risk Officers' Forum, a primarily European industry body focused on developing industry risk-management standards. Before joining ING in 2005, Dr. Wilson was managing director and global head of the finance and risk practice of Mercer Oliver Wyman, a strategy-consulting firm. Before joining Mercer Oliver Wyman in 2002, he was CFO and CRO of Swiss Re New Markets (SRNM), the alternative risk transfer and capital markets division of the Swiss Reinsurance Group. And before that, Dr. Wilson was principal and global head of the risk management practice at McKinsey & Company. Dr. Wilson received a BS in business administration from the University of California at Berkeley, and a Ph.D. in economics from Stanford University. Dr. Wilson has published extensively on risk management in professional journals and books.

About the Authors

Richard R. Lindsey is president and CEO of the Callcott Group, LLC, where he is responsible for directing research activities and advisory services. He is the Chairman of the International Association of Financial Engineers.

For eight years, Dr. Lindsey was president of Bear, Stearns Securities Corporation and a member of the Management Committee of The Bear Stearns Companies, Inc. Before joining Bear Stearns, Dr. Lindsey served as the Director of Market Regulation for the U.S. Securities and Exchange Commission (SEC) and as the Chief Economist of the SEC. He was a finance professor at the Yale School of Management before joining the SEC.

Dr. Lindsey has served on several corporate and not-for-profit boards, including those of the International Stock Exchange, Strike Technologies, New Hedge Fund Corporation, and the Options Clearing Corporation where he was the vice chairman.

Dr. Lindsey has done extensive work in the areas of market microstructure and the pricing of derivative securities. He has held the positions of Visiting Academic at the Nikko Research Institute in Tokyo, Japan, and Visiting Economist at the New York Stock Exchange. He has a BS in Chemical Engineering from Illinois Institute of Technology, an MS in Chemical Engineering from Berkeley, an MBA from the University of Dallas, and a PhD in Finance from the University of California, Berkeley.

Barry Schachter is Director of Quantitative Resources at Moore Capital Management, where he is responsible for risk management, financial engineering and trade analysis. He is on the Advisory Board and cochairs the Education Committee of the International Association of Financial Engineers. He is also a Fellow of the Program in Mathematics in Finance at the Courant Institute of New York University.

During the six years prior to joining Moore, Dr. Schachter served as Chief Risk Officer for three hedge funds. He also worked at Chase Manhattan Bank, where he was responsible for, among other things, the bank's market risk measurement models. He also was an Economist and acting Director of Research in the Economics Department of the Commodity Futures Trading Commission and an Economist in the Risk Analysis Division of the Office of the Comptroller of the Currency. He spent the early part of his career in academia, most of that time at Simon Fraser University and Tulane University.

Dr. Schachter has served as a member of the former Blue Ribbon Panel of the Professional Risk Managers International Association, and continues to be active with that organization. He has served as Editor-in-Chief of the *Journal of Risk*, and he is currently on the editorial boards of several professional journals, including the *Journal of Derivatives*.

He has been a regular contributor to *Risk* magazine, and has published his work in many academic and practitioner journals. His previous book was *Intelligent Hedge Fund Investing*. He also founded and maintains GloriaMundi.org, a website for risk managers.

Dr. Schachter received an MA and PhD in Economics from Cornell University, and a BS in Economics from Bentley College.

Index